SNMP
at the Edge

Building Effective Service Management Systems

Jonathan Saperia

McGraw-Hill
New York   Chicago   San Francisco   Lisbon
London   Madrid   Mexico City   Milan   New Delhi
San Juan   Seoul   Singapore   Sydney   Toronto

McGraw-Hill

A Division of The McGraw-Hill Companies

1 2 3 4 5 6 7 8 9 0   DOC/DOC   0 9 8 7 6 5 4 3 2

ISBN 0-07-139689-6

The sponsoring editor for this book was Marjorie Spencer, the editing supervisor was Daina Penikas, and the production supervisor was Sherri Souffrance. It was set in New Century Schoolbook by Patricia Wallenburg.

Printed and bound by R. R. Donnelley & Sons Company.

McGraw-Hill books are available at special quantity discounts to use as premiums and sales promotions, or for use in corporate training programs. For more information, please write to the Director of Special Sales, Professional Publishing, McGraw-Hill, Two Penn Plaza, New York, NY 10121-2298. Or contact your local bookstore.

 This book is printed on recycled, acid-free paper containing a minimum of 50 percent recycled, de-inked fiber.

*For David, who knows what it takes to complete
a writing project and whose daily understanding
and encouragement helped ensure the success of this one.*

CONTENTS

FOREWORD

Someone needs to help the network managers. Network management is broken.

Lore tells us that in March, 1987, a group of ARPANET geeks assembled in Monterey, California for a fine repast, the costs of which we can assume were passed along on company or government-contract expense reports wherever possible. For, over this meal, these packet-switching-powered amigos either discussed, dissed, or advanced such now-long-since forgotten constructs and notions as HEMS, SGMP, and CMOT. The point was to forge the precedents for a foundation of a management plane for the Internet Protocol (IP). IP was clearly becoming a pretty happening thing in the scientific research community and in bleeding edge enterprises. At that point nobody could have imagined IP being the impetus for doubling processor speeds every 10 months and unsolicited email advertising discount Viagra. Back then, "Spam" was still just a classic Monty Python sketch. The saga of Internet network management had begun.

Years of wrangling between European telco OSI sophisticates and IP ad-hoc working group trailer trash ensued in the standards groups. Emerging over several years was a little gem called the Simple Management Network Protocol (SNMP), which ripped a number of notions off from the high-calorie cryptic OSI/CMIP architecture and other network management architectures in the experimental phase.

SNMP had a number of innovations deviously buried in the simplicity of its design and framework:

A data-driven approach. It may be hard to believe now, but the notion of a management protocol that had a sparse number of fixed protocol operations and placed its richness in the database structure of the operands to those protocol operations was a pretty new approach. The idea of identifying network management scope as a database problem indicated a practical clarity that took a bit of getting used to.

An extensible naming scope for those data objects that comprised the stuff you could manage with SNMP. Compared to anything else labeled with the word "extensible" published in 1988, the notion of the Object Identifier (OID) has held up pretty well.

The very core of SNMP extensibility, the Management Information Base (MIB). This has provided a living framework to allow anybody with a text editor and patience for poring over RFCs defining the Structure of Managed Information (SMI), which define what managed data looks like, to enhance the very notion of the "stuff" SNMP can manage.

The result was that elementary management capabilities for entirely new protocols and notions on the network could be realized by someone who could focus on what they were managing (actual controls and indicators), rather than how they had to manage it (protocol operations).

I dwell on SNMP here because it can be seen as shorthand for "interoperable network management"—SNMP is the only framework that has held the promise for such interoperability then or since. Yet, fully thirteen years later, the idea of my being able to develop network management applications that can effectively manage (and for goodness sakes, even think of configuring) network elements developed by other vendors seems like it has advanced more slowly than the development of a formalized common language between me and my pet cats.

There are different reasons typically offered for this state of affairs by those who are SNMP advocates (hopeless romantic I am, I am still perplexedly among them), and those who have become SNMP loathers. The latter often include but are not limited to the poor saps who have been burned trying to make this stuff work in the last fifteen years in real networks, trying to solve problems of realistic scale using available tools. These reasons generally involve crummy implementations and weird market perceptions. And completely understandable market perceptions in the face of crummy implementations.

The fact is that in the cold light of day, secure interoperable network management, and the scalability and interoperability offered by the SNMPv3-era enhancements to the MIB framework hasn't seen the deployment success its inventors would have liked (to put it mildly).

To explore the reasons for this is to explore the nature of how interoperability is sacrificed. Let's start with what engineers in this space know well—there are few forces in nature more indomitable than the product manager at the network equipment vendor seeking to add value to his wares through proprietary and incompatible feature enhancements.

To be sure, a certain amount of thought and work goes into making the management and configuration of new features work in a heterogeneous management environment. However, a certain amount of work also goes into making the nonmangement aspects of a product operate in a heterogeneous network environment, such as carrying out basic functions like Dijkstra spanning tree calculation or route aggregation. An equal amount of invested sweat is necessary to demonstrate robustness in the face of oscillating convergence conditions, or in any of a number of real-world situational parameters that may not yield to a first pass of unit tests on a new layer 2 forwarding protocol on the Gizmo Gold router. Where does your equipment vendor feel comfortable compromising on drawing that line?

If the compromise is, as it often is, in the control and management plane, the ad-hoc solution is frequently found in pouring a pint of engineering into the proprietary management interface rather than a quart into the industry-standard one (aka, SNMP). For modern network elements, this translates into ye olde Command Line Interface (CLI). Now, admittedly, however a CLI shows up for a vendor, it solves a number of problems SNMP has failed to up to now. One is presentation in a profoundly human-intelligible form (assuming the human has sufficient domain expertise and has put the up-front effort into learning and absorbing the *Kama Sutra* of the specific vendor CLI). Another is adaptability to automated scripting.

However, the greater management picture isn't covered by the CLI; it generally can't be.

The CLI usage case envisioned by that product manager we talked about earlier is the single human operator typing at a CLI for a specific vendor's device. That won't cover a scalable deployment of many customer-provisioning changes percolating from the network edge on in. It won't cover dispatch of unsolicited notifications to allow fault recovery in somewhere close to real time. Only a protocol and framework mechanisms proved robust under demands like this, engineered with the techniques to guarantee that robustness, will make the industry grade.

This brings us to the quality of the SNMP agent implementation. Is there one at all, that covers that new layer 2 forwarding protocol? For that matter, how much energy has the vendor put into the SNMPv3 features that allow secure configuration?

See, it isn't as though the keepers of the flame of SNMP-based management have been napping during this entire time and failing to notice the problems of modern service deployment into enterprise, provider, and carrier environments. Jon Saperia has been one of the continuing pioneers in this area, and the sophistication of the implementation and deployment approaches that he provides you in this book are sufficient illustration of all that has been going on in this area. For whatever reason, network equipment vendors just not "getting it" as to the importance of offering interoperable, scalable management of services and not just their latest features. The IETF as a standards organization has to shoulder some of the blame here too, in failing to clearly identify the customer value proposition of the framework innovations. Stated differently, I can identify some of what their lack of solving the disconnect between what's out there and what's possible has cost the intended product constituency.

For starters, the proliferation of the proprietary in the world of network management tools is ridiculous. It seems that there are two kinds of network equipment vendors in the world—those who accumulate individual network element managers (that don't work together), and those that lost their collective shirts amassing all of their development resources in a single element manager (which generally, never worked at all). Can it really be that hard?

And as for interoperability? If I had to wager on adherence to common semantics across any arbitrary ten network element vendors among each other, I wouldn't bet on being able to write a seamlessly interoperable management application to access more than a few paltry tables dating from the original MIB-II. While that may be enough to let me know some pretty fundamental things (does the interface appear to be working?), it's also sobering that these are tables from the MIB which was the original proof of concept of SNMP all those years ago. It's only a bit hyperbolic to suggest that if the telephone had followed a usage and development history like that of SNMP-based network management systems, by 1940 people would have been using phones primarily to call each other to exclaim, "Watson, come here, I need you."

So, hats off to the network managers who have been able to demonstrate value out of all of this in the jungle of operational reality. The state of many of the SNMP management tools they use has not changed substantially since when The Lambada was a new thing on the dance scene, when Milli Vanilli were at the ascendancy of their pop careers, and a 25 Mhz 80386 was a "hot" personal computer CPU. It predates by several years a presentation to a skeptical IETF audience by a soft-spoken Swiss physicist named Tim Berners-Lee of a new information retrieval system he had called the World Wide Web. And it certainly predates the technologies prevalently deployed in modern carrier and enterprise networks. Now, this may partially speak to the endurance of the underlying SNMP architecture and constructs. Let's face it; people have done amazingly innovative things with fairly basic managed element instance data. But, it is also an indictment of how the technology has been adapted (or not) in new products, to create new tools, to meet new challenges.

The result is a set of expectations at the lowest rung, which when it comes to product management and engineering, always seems to be a self-fulfilling proposition. The world of flawed implementations has brought the customer base to believe in a flawed technology. You can't blame them, really—given that they're out to solve problems and generate revenue, the difference isn't really meaningful.

I first ran across Jon Saperia in the late time sharing age of 1983, when we were both at minicomputer leviathan Wang Laboratories. Fortunately for me, our paths have crossed a number of times since. He has compiled a storied career as being one of the most senior architects in the networking group at Digital Equipment, and then as a Software Development Director at Ironbridge Networks after that, and a bunch of stops in between. However, my primary awareness of Jon over the years has been in his unique contributions to the advancement of network management in the IETF.

From his earlier work specifying the SNMP management capabilities of the notoriously complex DECnet Phase IV protocol architecture, Jon has provided a perspective remarkably lacking over the years in the IETF network management activities: the perspective of a world-class network application engineer who understands not only the needs of network management applications, but also how to manage network-based applications.

Most recently, it has been my unique pleasure to cooperate with Jon in the SNMP Configuration Working group of the IETF, and in particular, collaborating in coauthorship of a Best Current Practices document on Configuring Networks and Services using SNMP. In all of the years I have followed SNMP, it has never ceased to amaze me how secondary the goal of using it for configuration has become, and I have always felt the ubiquitous advancement of SNMP to this application is essential to the success of its goals. In my collaboration with Jon, and our other gifted coauthors Michael MacFaden and David Partain, I have found kindred spirits in that belief.

Now, it is great that there are keen minds in the network management area of the IETF who can address issues of byte ordering and SMI esoterica, God love 'em all, I don't know what we'd do without 'em. I am blessed to count many of them as good friends. However, Jon (who is certainly no slouch in the area of the protocol and MIB specifics in his own right) has an approach that prioritizes solving the problems that nobody else seems to be taking on, but that clearly comprise the substance of what concerns the people deploying network management systems, that must solve a class of new problems, right now. What Jon uniquely understands is that the difference between management applications blissfully counting deltas in ifOctetsOut and solving the real higher order problems facing people deploying network services is like the difference between ability to demonstrate basic motor skills and ability to execute a NBA full-court defense for four quarters. This is to say, one can be inherent in the other, but it doesn't necessarily lead to it. Higher-level thinking, abstractions, and practices are required.

SNMP at the Edge looks to identify the elements of that thinking, to find those abstractions, and to identify and even invent those techniques. It is part of a very small class of new books on network management that seeks solutions to make network management not just the stuff of solving immediate problems in the disruption of the data paths of application service traffic, but to use the services of network management to in fact enable a new class of network application.

Jon coherently defines and constrains his key principles and elements in this application space—the service, the policy that drives the configuration and management interactions with the service—and surveys the dynamics governing the service management application. This isn't a book just about network management, and certainly is more than a book about SNMP. It's about a new way to think about applications that interface with network elements and processes which run on them, a way of thinking powered by identifying new opportunities for generating revenue, and providing products and services (in the broader sense) to customers and in-house enterprise constituencies. However, along the way, Jon thoroughly illustrates a higher order way of looking at and using the mainsprings, screws, techniques, and developments of network management, particularly SNMP as it has been creatively evolved in the last seven years or so.

If you're holding this book right now, there's a good chance you already have a good understanding of the problems at hand. If you have bought this book, then you have at your disposal the tools and techniques to help go out there and solve the problems at hand. So for the benefit of us all, get out there and fix the state of network management, will ya?

Wayne Tackabury
Engineering Manager, Router Configuration Group
Gold Wire Technology
Martha's Vineyard, Massachusetts
May 2002

INTRODUCTION

With the fall of the dotcoms and the general economic slow-down that accompanied their demise, many service providers found that revenues and margins were not going to grow they way they hoped. Which was cause and which effect—supposing they're not independent phenomena—doesn't really matter. Service providers are looking for services they can sell that are more profitable than raw bandwidth—which has become a commodity. Network infrastructures become more important for service providers and enterprises as they realize that services layered on top of their valuable infrastructures can improve productivity and reduce operational costs.

The logic seems so simple, the proposition so straightforward, that many were surprised when the new services failed to materialize. If everyone wants to make more money and reduce costs, where are the products? Of course there's the traditional chicken-and-egg problem: customers seldom ask for new services since service providers do not offer them. Even if customers do ask for new services, most service providers still do not offer them for a number of reasons, one of which is a primary focus of this book—the cost of service creation and management.

This cost, at least in part, is driven by the tools available to network operators for creating, deploying, and managing network services. The management techniques most vendors products support have changed relatively little in the past 10 years—astonishing if you think of how the Internet has grown and changed during that same period. Through all of that upheaval, we have continued to configure and manage our IP network infrastructure as we always had. To be sure, we now have better SNMP-based tools and many network operators have developed sophisticated scripts that interact with Command Line Interfaces (CLI). But the basics remain the same: configure with a script or the protocol and data type du jour, do some SNMP polling, and hope for the best. Small wonder we're so tentative about the creation of high value services and are at a loss about how to bill for them accurately.

An inhibiting factor for more substantial improvements is that each year at least one new holy grail of management emerges. Unfortunately these seem holy only to the people who created them—the protocol engineers. As I've often said in my speaking engagements:

—What do you get when you lock a group of protocol engineers in a room?
—A new protocol.

The fact is that even if new protocols were needed, they would not really solve much of the problem because each protocol tends to focus on a narrow problem area.

Successful service creation, provisioning, management, and billing are not separate problems, they're an integrated problem. It has little to do with management protocols and quite a lot to do with our failure to understand the full scope of the job. That scope incorporates:

* The customer's view of the services he relies on.
* The network topology at many levels from the physical layer to the applications layer.
* Each of the technologies needed for service creation.
* The standard aspects of these technologies and how the vendors "improve" them.
* The means of configuring and managing these technologies across an increasingly heterogeneous network infrastructure.
* Billing and service-level issues.
* Infrastructure requirements for every service—for example, a mail system that requires a functioning name server and a Web service will usually also have the same requirement for a reliable name service even though they are different types of services.

This is all far too much for mortals to keep in their head; the very few that can are as scarce as hen's teeth, and expensive. More expensive, in several senses of the word, than the management tools that might make us less dependent on them.

What This Book Is About

SNMP at the Edge is about service management, particularly the development of management software for profitable service offerings. I use the Simple Network Management Protocol (SNMP) as a way to explain the problem because it is so widely deployed and provides an excellent foundation on which to build service management and verification software. Note that I say "foundation," not "solution," because no protocol or framework is the whole of the problem. As we will see in later pages, the greatest technical difficulties for service management lie in the development of the management application software. The problems that confront us in that arena are not protocol-specific, although the choice of protocols used to communicate with managed devices does indeed impact management software.

In Part 1 of this book, we examine types of network services from simple name servers to complex videoconferencing. Services are described independently of any specific technology or protocol that could be used to configure and monitor them. Any system designed to manage such services—a service management system—must serve several different constituencies. That fact alone means that operators have a number of different perspectives from which they might manage a service, some quite alien to them. Part 1 reviews service management requirements in the context of these disjoint perspectives, and continues with an examination of the different types of edges in a network. An edge is created any time we encounter an administrative boundary, regardless of the type or capacity of the equipment. Edges exist both within and between organizations. This understanding of services and the special challenges of management at the edge are fundamental to developing software and hardware that can offer cost-effective services.

Policy and policy-based management have received a lot of attention over the past few years. Despite the publicity, there's still a lot of confusion about this approach to management. Part 2 delves into concepts in policy-based management that are relevant to service management systems and suggests how those concepts can be used to define management system requirements. Because some of the available technologies overlap and can make design and deployment more complex, we review some of them to identify which are of most help. Part 2 concludes with the creation of a generic network model of services based on simple object-oriented principles.

In Part 3, using the groundwork laid in the preceding chapters, we'll construct a design for a workable service management system. This discussion begins with a review of the requirements for a base technology and how SNMP can be used to satisfy them. While not perfect, SNMP is the best available choice for a comprehensive service management system. Current work to expand the configuration capabilities of SNMP is reviewed, not just as a simple matter of information, but because this work incorporates some of the important principles of policy-based management into the SNMP environment.

Part 4 proposes a design for management software that will run inside network devices such as servers and routers. It concludes with a close-up look at how a well-designed management infrastructure in a managed device can be used to support a service management system. Existing MIB objects are used to show how important services could be configured and monitored to aid in service level reporting.

Part 5 begins with a discussion of special-purpose management software that can be developed for managed devices that will further enhance the ability of a service management system to efficiently perform its tasks. Some of the techniques include development of software that provides high-quality information to management systems when there are operational or configuration difficulties. Chapter 11 proposes an approach to building an effective service management system that can take advantage of all the features proposed for managed devices and with this foundation laid, Chapter 12 shows how the different parts of a complete service management system might interact with each other.

After a brief review of external interfaces for a service management system, the book concludes with a look into what the future might hold for service management software and the role vendors, customers, and standards bodies play in shaping that future.

Who This Book Is For

SNMP at the Edge is for people interested in the design, deployment, management, and accounting of value-added services in IP-based networks. It is especially directed at those with an interest in the software that controls and enables these valued-added services, be they voice over IP (VoIP), business-to-business (B2B) videoconferencing, or very secure virtual private networks (VPNs). In particular, readers who work for service providers will find the material interesting, if not always consistent with their views. (A service provider in this context is any organization providing high-value services and is not restricted to Internet Service Providers [ISPs].) Many enterprises with large-scale networks face the same challenges as do ISPs. In fact, many of the problems that enterprises know all too well—integrated management of networks and hosting centers—are now becoming significant concerns for ISPs too. If you run a network, there is probably something in the following pages to challenge your assumptions.

This book describes many network and system management tools and techniques that are not yet available from suppliers of network equipment and software. As a result, many people who work for vendors of such systems will also be interested in much of the material. A significant portion of these pages is devoted to the design and implementation of hardware and software systems for the type of service management at the edge that can make layered services a reality. While the Simple Network Management Protocol and its various extensions in a number of IETF working groups are used as the example technology for much of the discussion, readers will find it easy to extract general design principles and apply them using other technologies.

This book is not a tutorial on basic SNMP and some background will be helpful in those sections that deal with SNMP-specific technology. For example, I'll discuss the protocol and MIB Structure insofar as they can be used as foundational components of an integrated service management system, but will not detail the format of each SNMP protocol data unit. Neither is this book intended to be an introduction to database or object-oriented design, but some basic concepts of each are invoked for the discussion. This very breadth of knowledge areas required for the informed development of management software is one of the problems associated with the creation of effective service management software. So many skills are needed to develop network management software and the more complex service management systems that practitioners tend to become specialists in one area. They are specialists in management protocols, database technologies, user interfaces of various types including command line interfaces, and many other disciplines. While these disciplines (and more) are essential, we sometimes get lost in the details and forget about the global problem that we are all attempting to address—software that enables the cost-effective deployment of high-value services. *SNMP at the Edge* offers a wide perspective. If you have a good background in any of the technologies needed to build a management system, then your appreciation of some sections will be enhanced. If, like many, you are not familiar with a lot of the topics in this book, you will come away with some insight into the problem area of management software in general and one comprehensive way of defining the problems of network management software in particular.

There are usually several productive ways to approach a complex problem, and services management has proven to be complex. This book presents one view—my view—of how best to approach solutions for supporting profitable services in an IP network. It does not presume to be *the* approach. The Internet Standard Management Framework, SNMP, is positioned as the primary technology for purposes of illustrating what an effective service management system might look like. Work to improve SNMP technology is ongoing; just as work to combine and expand the capabilities of routing is. My choice of SNMP is based on the time-honored principle of keeping the number of protocols and technologies in our management infrastructure to a minimum. It is a conscious decision that a general tool, while imperfect in many respects, is better than a collection of uncoordinated technologies, some performing configuration, others fault management, and still others accounting functions.

Even though SNMP is a core technology used here for implementation, the concepts are transferable. If your design choices lead you in other directions, feel free to add the governing principles of this analysis to your own understanding of the problem, and apply them toward new solutions using your favorite technologies.

Learning More

SNMP at the Edge makes references throughout to the Internet Engineering Task Force [IETF] and its various working groups and their publications. It also references other organizations. You'll find an appendix on useful URLs at the end the reference section to aid further research. It's needed because some of the work by these organizations is continuing, and individual documents referenced here may be valid for a relatively short period of time as they are superseded by new publications.

ACKNOWLEDGMENTS

Much of my thinking has been influenced over the years through my work in various Internet Engineering Task Force (IETF) working groups having served in a number of roles including chair or co-chair, working group member, technical advisor, and draft author or co-author. In each case, I have benefited from the counsel and sometimes animated interactions I have had with the working group participants. The IETF has contributed much to my theoretical training and understanding of network protocols and behavior over the years.

Many of my co-workers, especially those who have worked with me on management software development, have greatly informed my views about the best ways to go about the development of management software whether for sophisticated applications or for inclusion in managed devices. Lastly, I would like to acknowledge the contribution my customers have made to my understanding of the problems discussed in this book, especially those who design and manage networks, since they have the most difficult task of all, making the network work.

There are too many individuals in each of these groups to list, so I would like to thank everyone who has participated in my "education" over the years. You know who you are.

I would like to offer a special thanks to those who have invested their time discussing this book, looking at early versions of the manuscript, or offering advice and comments: Bruce Boardman, Pablo Halpern, Wayne Tackabury, and Bert Wijnen. A special thanks also go to Marjorie Spencer at McGraw-Hill whose patience and care helped to bring this project to fruition.

Service Management

Before we can discuss new concepts that will enable the creation of effective service management systems, or the design principles used in their creation, we must first define and examine in some detail what we mean by *network services*. Before we can propose an approach to building a service management system, we must first identify the problems that it will be constructed to address. Part 1 lays this foundation for the rest of the book.

- Chapter 1 examines network services and describes how services can be viewed from a number of different perspectives, including that of the provider of the service as well as the user or consumer of the service.
- Chapter 2 defines a system designed to manage these services, and identifies the functions it must perform for effective service delivery. It also analyzes some of the factors that have inhibited our ability to create such systems.
- The edge—loosely defined as the administrative boundaries between organizations—presents significant challenges to service management software. Chapter 3 focuses on the special issues confronting service providers and users at the edge, and how these issues affect service management software.

CHAPTER 1

What Is
a Service?

We begin with an examination of network services and what we mean by them. We take this approach rather than immediately plunging into the issues related to the design and development of service management software because how we view these services has a significant impact on our thinking about how to design and implement service management software.

Services in a modern network can be divided into two broad categories: *visible* to the customer and *transparent* to the customer. Visible services, such as videoconferencing, might rely on a number of invisible services, such as routing and name services. One problem blocking profitable service deployment is that there are basic services—absolutely necessary for proper network operation—for which most people are not prepared to pay very much. By and large these infrastructure services are transparent by nature. Without them, however, more sophisticated services—*high-value customer services*—are not possible. In our video-conferencing example, a high-value service used in the delivery of the videoconferencing product might be quality of service (QoS) to ensure that the picture at each end of the conference is as stable as possible. Unless the routing, name service, and basic infrastructure components are all operating effectively, there is little chance that the QoS functions will work correctly.

Infrastructure Services

Infrastructure services are those that the basic system needs to operate. Let's use the telephone system as an analogy. We take basic calling services and the transmission infrastructure between our homes and businesses for granted. Over this basic infrastructure other services like caller ID or call forwarding are layered. These high-value services depend on the transmission and basic telephone infrastructure. When an infrastructure service fails, we notice it, in part as a result of the other services that disappear with it—the most basic of which is the telephone connection. Although we know that failures can be caused by a storm, an equipment breakdown, or human error, these systems have become so reliable (in most parts of the world) that typical users do not even think about a special service to guarantee their uptime.

Nonetheless, in some cases we value connection so highly that any outage is unacceptable. In these cases we are willing to pay for a higher level of service. We purchase a guarantee of continued function even when some parts of the network fail—a *service-level agreement.* We will talk about the issues related to service level agreements later, but for now keep in mind that some basic telephone services are so important that people will pay extra to have redundant service to avoid outages. The same is true of power utilities. Some people consider the continuous availability of power so critical to their enterprise that they invest in costly backup systems. There are two interesting points to absorb here:

- Infrastructure services are the foundation on which more lucrative customer services can be based.
- Some infrastructure services are so valuable in their own right that people will pay a premium for high availability.

How many of us settle for basic telephone service? I suspect not many. Most of us have a longish list of add-on services that have become so common we can hardly imagine a telephone without them: call waiting, caller ID, call forwarding, voice messaging, and more. These add-ons, or *high-value customer services*, are layered on top of the basic infrastructure services.

The same is also true for cable television service. Like many customers, I started with basic channels and maybe one or two other extras. A comparison of my cable bill with one many years old reveals a staggering increase, due in large measure to added-value services called "premium channels." It turns out that they are packaged in such a way that I often agree to take some services I neither want nor need to obtain what I do want. The cable companies have learned what many car manufacturers already know: to simplify production and distribution, limit the customer's number of choices.

From a data network perspective, there are both infrastructure and high-value customer services as well. The idea of basic services that have other, more sophisticated services associated with them is an important principle that we will use when discussing the creation of service management software. While your mileage may vary, here is a list of services that are prerequisite to building other revenue-generating services:

- **Basic network connectivity**—The leased lines and interconnecting infrastructure (much like the telephone or electric transmission system).
- **Routing**—The service that still consumes the vast majority of brain power among the technical specialists running IP environments.

- **The Domain Name System [DNS]**—Commonly used to convert human-friendly names such as www.foo.com, to IP addresses used by network systems.
- **Electronic mail**—So basic that it, along with access to the Web, constitutes the "dial tone" of the Internet.
- **Basic Web services**—Many service providers bundle some rudimentary Web services in their entry packages, although the disk space available to customers is small and there are seldom any service level guarantees. Most ISPs do offer a "home page" for their subscribers, who can use it as a portal to other Web servers on the network. Basic Web services also help generate at least some revenue for the service provider by inciting users to create their own home pages, which sometimes are sold at an extra cost.

Note that even among infrastructure services, there are dependencies and interrelationships. For instance, the e-mail and DNS services don't function without basic network connectivity and routing. Also keep in mind that although these services are commonly provided by ISPs, some of them could be provided by the business customer. For example, some businesses run their own electronic mail and name servers: the exact location of the servers and who runs them is not important, only that they are present. E-mail and name servers are prerequisite building blocks for other more advanced services. In the next section, we examine what happens when even more complex services are layered on top of the first, and we investigate some of the more important interrelationships.

High-Value Customer Services

High-value customer services are any combination of infrastructure services that can be deployed profitably. *Profitability* implies sufficient usage of the service, configured in a similar fashion, to make management somewhat tractable. It is possible to link one or more of these high-value services to create an aggregate service that expands usage for each of them.

In a sense, the World Wide Web did for the Internet what the spreadsheet did for the personal computer (PC). Just as PCs have become commodities in many respects, so too have Internet services. The difference is that many, but not all, PC manufacturers have found a way to sell these commodities profitably. Just as many of these PC vendors did not understand at first how to be profitable, network service providers have yet to learn new ways to increase their profitability. One way to do this is by selling accessories and add-on services. This is particularly true for the larger companies, especially those that cater to businesses. Service providers, whether they are enterprises or ISPs, are seeking ways to get added value out of their network infrastructure investments. Indeed, part of what drives the proliferation of application service providers (ASPs), ISPs, and other service provider types, is the desire to create lucrative services out of network building blocks, either alone or in combination with new services. For a service provider to offer application or databases services, it must either team up with another service provider or supply the infrastructure services on which the applications depend. Sometimes new service providers are created to focus on a specific type of service, such as application hosting. In other cases, new organizations within existing service providers are created.

Network operators seek the equivalent of the PC's spreadsheet. Consider the spreadsheet: not only was it a valuable tool in its own right but, when combined with the other functions emerging on the PC platform, it created a critical mass of capability that spurred PC sales. Prior to the PC, dedicated word processing systems were selling quite well. The spreadsheet was an application that rapidly put the word processors out of business, even though many word processing vendors did try to fight their way back by putting spreadsheet programs on their systems. Why didn't their strategy work? The decisive emergence of the PC was driven in large measure by the rich range of third-party applications that could run on standardized PC platforms. It was the combination of the utility of spreadsheets with other applications (the analog of a service in the network) that helped to drive PC sales. (In a sense, the World Wide Web was the first killer application for the Internet.)

What are these new high-value services for the Net and how are they related to each other? Because these services do not yet exist, at least in a form that is widely accepted and recognized, it is difficult to discuss their utility or how one would attempt to create a management system for them. The implication is that service management systems must be flexible, otherwise we end up writing a customized application for each new service that is created. Later in this chapter, we'll taker a closer look at the limiting factors constraining service availability today; for now we can talk in terms of some services commonly believed to have high value, many of which appear to hold the promise of increased value to users when combined in different ways.

Let's start by pointing out what these high-value services are not. They are not enabling technologies. Enabling technologies such as Differentiated Services [DIFFSERV], Integrated Services [INTSERV], and Mobile IP can form the foundation for more interesting services.

For the present discussion we are not interested in the Differentiated and Integrated Services technologies per se, but we are interested in what can be done with them to create a service that customers will value. Differentiated and Integrated Services are technologies that can be used to guarantee that certain types of traffic between systems receive the resources they need to meet minimum performance requirements. For example, Differentiated Services could be used to deliver guaranteed performance to a videoconference site for a customer. From the customer perspective, the high-value service is the videoconferencing itself. As we shall see shortly, it is likely that more than one infrastructure service will generally be required before we can deliver a service such a videoconferencing. Mobile IP is another technology that is not a service the customer directly values; however, Mobile IP technology deployed so that customers can get electronic mail and other services while away from their desks is something that some people value.

Although one admits to ignorance in choosing what might be the next service breakthrough from a provider's perspective, there are some services that people tend to gravitate to when the discussion turns to high-value services. These include:

- **Virtual Private Networks (VPNs)**—This is not the same thing as the IP Security Protocol [IPSEC]. It is not the IPSec technology per se that is of interest—what *is* valued is a service that customers can purchase to allow the secure transmission of their confidential information. Such a service may *use* IPSec, probably in combination with other network infrastructure services, all of which are transparent to the customer. Many enterprises are interested in VPN services because they want to use third-party network service providers instead of shouldering the cost and hassle of managing their own networks. They may currently use expensive leased lines from service providers to help assure a reasonable level of security and performance for their communications. If VPNs are deployed as a reliable service by service providers, these enterprises could purchase the service and eliminate at least part of the need for their own high-cost leased lines. For those companies, the cost of VPN service will seem quite reasonable by comparison.

- **Voice over IP (VoIP)**—This is actually a borderline case. The real value in VoIP may reside more in the integration of the voice function with the rest of the data network—and everything that implies—as opposed to any inherent benefit in transmitting voice via IP packets.
- **Video teleconferencing**—This is a genuine candidate for a service to be layered on top of a network infrastructure, and it has the potential to be highly valued by at least one set of customers, enterprises in particular.

Aggregated Services

Aggregated services are combinations of any number of infrastructure and high-value customer services. Think of aggregated services as a sort of "business application." For example, consider an aggregated service that delivers secure Web pages to users on a network. It might be built out of several infrastructure and high-value service components. The infrastructure components could include basic network connectivity, routing, and DNS services.

The high-value services might include a service that guarantees that the traffic on the secure Web server is treated with a high priority. This could be accomplished with Differentiated Services or any one of a number of other technologies, and a service that creates a VPN between the Web server site and certain other locations for added security.

From a user perspective, all of the features must be working correctly both individually *and in concert with each other* for the system to work. Hence, in our simple example, not only must the basic network infrastructure be working, but also the routing and name services that support the Web server, and all the network elements between the server and client. In addition, the elements involved in the delivery of the high-value services that support this user-visible service must be configured so that the desired traffic not only gets to the right place, but also gets there within the time frame guaranteed, with the appropriate level of security. If one piece—say, the service that guarantees performance—fails, then we could say that the entire user-visible service has failed or is not operating at satisfactory levels. One can only determine this by understanding the relationships between the services that have been combined to create this aggregate. These often-complex interrelationships make service creation, management, and accounting difficult.

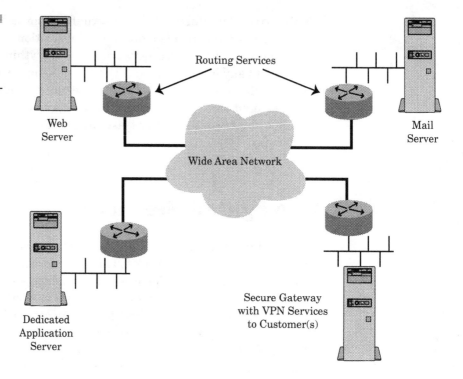

Figure 1.1 contains examples of infrastructure and high-value customer services. The former is represented by the wide area network (WAN) infrastructure that consists of T1 and higher-speed lines and the interconnecting routers that provide the routing services. (A real network would be much more complex than this one, with redundant links, etc.) High-value customer services are represented by the secure VPN gateway. As we've noted, e-mail and Web services are likely to be fairly basic offerings, but in our example, they are aggregated to create a service that is much more sophisticated than the Web and mail services people normally have.

Using these building blocks, it is possible to imagine an aggregated service that could be offered profitably and have a demonstrable high value. Imagine we have the Acme Online Products Company (Acmeon) that sells products over the Web. To do so, they purchase Web and other services from ISPs. This model will be familiar to anyone who has purchased almost anything via the Web, but the important features of this service from Acmeon's perspective are:

- Available Web pages have sufficient responsiveness to meet the demands of their customers, even during peak periods of demand. This facility is provided by the Web server in our diagram.
- The Web server is tied to the application server in our diagram. This connection is necessary so that customers can be informed when an item is out of stock and will cause a delay in their shipment.
- Mail confirmation lets customers know when the order is placed, when it is shipped, and any difficulties that may arise with shipment or delivery. In our diagram, the mail server is connected to the Web and application servers so that they can tell the mail server what types of messages to send.
- Secure internal communications within Acmeon to specific sites, as well as secure communications with Acmeon suppliers, are used for automatic inventory replacement and other functions.

Having itemized the service features, note that the real power of the service is in the integrated operation of all of these components. Not only must components such as the Web and mail servers work correctly by themselves, but they must work correctly as an integrated whole. This is the high-margin value of an aggregated service.

Next, let's describe the service from several perspectives: Those of the network service provider, the service provider's customer (Acmeon), Acmeon's suppliers, and Acmeon's customers.

Service Perspectives

As we can see from the Acmeon example, the services outlined can represent a higher level of complexity than many service providers can comfortably juggle at the current time. Adding to this technical complexity are complexities introduced by the subjective concept of "value." Different interested parties may have different perspectives on this same service; in the example above, we described the service provider and its customer (i.e., the organization that pays the provider for the service). Whereas Acmeon is the primary customer, the people who actually produce or supply products for the online retailer to sell act as suppliers to Acmeon. Acmeon's profits come from their customers who are, from the service provider's perspective, the customers of their customer.

The concept of *service perspective* is important, because many of the problems associated with service creation and management are relevant to some perspectives but not others. The four perspectives to be elaborated in this section include:

- The customer's perspective—the retailer
- The customer's customer perspective—the people who order retail products
- The customer's supplier's perspective—the people who supply the retailer with inventory
- The operator's perspective—the provider of network services to the retailer

There are any number of other perspectives within a service organization that may be useful, but the important point is not so much the details of each of these perspectives, but that we must understand how they interact to deliver high-value services.

Our goal for now is to see how different perspectives affect how one understands and defines a service. Later we will see how these perspectives impact our thinking on the design of service management software.

Customer Perspective

Customers such as Acmeon predictably have a mental picture of what they need from their provider. Acmeon wants its customers to be able to order products securely over the Web site, and receive immediate email confirmations about order status. This Web site is also connected to the inventory management system, which in turn is tied to Acmeon's suppliers via a VPN. This VPN therefore enables both the retailer and its customers to place orders and check order status. Of course, Acmeon's customers know nothing about the VPN, which works transparently, behind the scenes. Additionally, Acmeon very likely runs its inventory management applications on its dedicated application server in the operator network as part of the service.

Implicit in the definition of "service" in our example is the concept of relative priorities about different types of traffic in the network. For example, the provider's customer may have several unsecured Web pages on its Web site, as well as a set with Transport Layer Security [TLS] for Acmeon customers who place orders using their credit cards. When customers transmit credit card numbers, they feel more comfortable knowing the information is secured to the appropriate degree, even though the majority of them will not understand the details of how the security is provided. Acmeon might, in fact, want to pay extra to ensure that the relevant order-taking pages always get very fast performance: a waiting customer is a frustrated customer, capable of taking his order elsewhere.

The essential point here from Acmeon's perspective is that there will be increased cost associated with Web pages that are secured versus those that are not. And because it wants quick turnaround for these pages, secure pages may also get expedited service in the overall system. Both speed and security are likely to cost Acmeon money, so it will wish to treat only essential information with this particular set of services. Acmeon will want to validate these services, not only to ensure that it gets what it pays for, but also to ensure that customer information is secure.

A number of other possible aspects to a customer perspective will become interesting for later service creation. For example, Acmeon would probably be very interested in accounting and billing information for the service, as well as current and historic service status details.

Figure 1.2
A service customer's perspective—Acme Online Company.

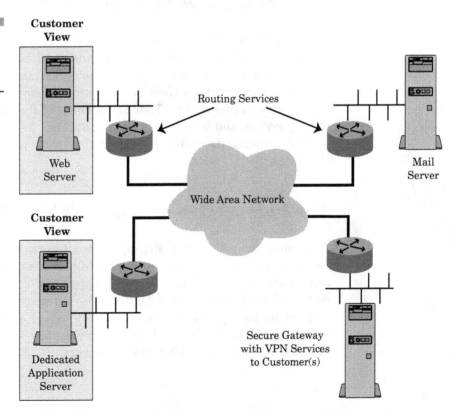

Note that Figure 1.2 does not highlight VPN services in the same way as application and Web services. The VPN might be used to connect Acmeon's corporate offices with each other or to connect with selected suppliers for the exchange of sensitive information. To some degree, Acmeon, like any customer, assumes the VPN and infrastructure services, (routing, DNS, etc.), are present. These services only become really noticeable when they fail; the service customer sees only the performance and reliability of the Web and application servers. From a customer perspective the email, VPN, and infrastructure services are important, but, like telephone transmission for most of us, they are transparent. When discussing Acmeon's customer's perspective, email services (for example) are important because that is how a customer tracks the status of his order. But since Acmeon's order-tracking systems are automated, the e-mail they generate is transparent to the retailer and hence not highlighted here.

The Perspective of Acmeon's Customers

From the perspective of Acmeon's customers we have a simplified landscape. I've highlighted only the Web and mail servers in Figure 1.3, because Acmeon's customers' view of the system is the Web server (i.e., how well it responds to their orders) and the email notifications they receive about order status. They see nothing of the back-end order processing system and the virtual private network.

Like Acmeon, its customers expect the rest of the infrastructure to "just work."

The Perspective of Acmeon's Suppliers

A supplier has yet another slightly different view. It will probably want Acmeon's dedicated application server to query its inventory for out-of-stock items directly. Likewise, the supplier might send updates to the dedicated application server. On the other hand, the supplier may have little or no use for Web pages or the automated email system. It may interact with Acmeon via a Java application, or with the inventory and ordering system via the VPN in the diagram.

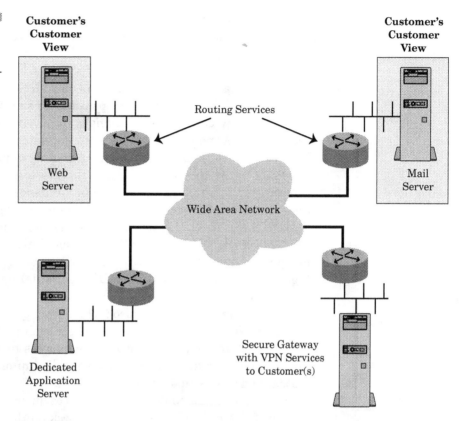

Figure 1.3
Acmeon's customers'
perspective.

The Network Operator's Perspective

The network operator has the most complex perspective. In fact, the operator has several perspectives, because it is likely that several organizations within the service provider are involved in the delivery of such a service. What the service provider employee sees of the service depends on which department he is in. Even within a technical area, several departments may be involved. Typically there are people who focus on the raw network infrastructure, such as the transmission lines that connect the routers and are represented by the WAN in the previous diagram. In many organizations, these people are distinct from the group who provisions the routers. These two groups are often separate from the people who take care of customer servers, who in turn staff different groups from those who run the DNS and mail servers. This is perhaps an extreme example, but variations on the theme are common. We will examine the implications of this "distributed" approach later, but for now, an integrated view of the technical requirements for Acmeon's service looks like this:

- The secure gateway to support Acmeon's virtual private networks must be configured. The group responsible for setting up the VPN will also make any configuration changes necessary to allow suppliers access to the Acmeon network when appropriate.
- The routing group may want to make routing adjustments to create alternate paths through the infrastructure to ensure that Acmeon gets the best possible reliability and performance, from their perspective. That way, if there is an outage to its main site, Acmeon can continue to access its application servers and monitor its business.
- The ISP offers DNS and hosting for companies that purchase our imaginary high-value service. For each customer who takes advantage of the offer, the provider's DNS server(s) must be updated as well.
- The service we're describing also includes two types of electronic mail functions. The first is a simple one where mail addressed to Acmeon is processed and made available for employees to pick up with their favorite mail agents. The second type is the automatic order confirmation and status updates triggered by activities on the application server. If one of Acmeon's customers places an order on the Web server, and the item is found to be out of stock on the application server, there is a further technical requirement to modify the mail server so that it can notify the customer and, quite possibly, accept changes to the original order.
- The Web server provides a front end for this order processing application, which may be quite large in itself, and additionally connects to inventory and other large databases. The Web server also provides the more basic service of hosting the company's home pages.
- Implicit in our description of the service that Acmeon has purchased from the ISP is some guarantee of service-level quality. People who access the secure page to pay for products get service that is guaranteed to be at a minimum level of responsiveness for those portions of the service under the provider's control, no matter how busy the rest of the system is. To achieve this result, it may be necessary to adjust parameters at the edge devices,* as well as at the servers. Acmeon is willing to pay its ISP for the service guarantee, and in turn the ISP may have to commit more resources to deliver what the customer requires. Under some conditions, the ISP may even have to pay a

penalty to its customers if it fails meet the minimum agreed levels of service.

✳ A great deal of data related to the operation of elements in the service provider network must be collected. These data are used for service level verification and accounting. Without this information it is not possible for the ISP to demonstrate to Acmeon that the service purchased has been delivered as promised.

No wonder so few of these complex services are offered. To make all service systems run correctly and guarantee that they will stay that way is a real challenge—one beyond the capability of existing management technologies. This sheds some light on the question of why we don't yet have a menu of high-valued customer services: to a large extent, our ability to deploy and accurately invoice these new services profitably is limited by the current state of management software. In the next chapter we describe the requirements for a new generation of service management software and investigate the issues that inhibit its creation.

---

*Service providers may have to develop more sophisticated relationships with each other to support some high-value services so that quality does not vary too widely as traffic moves through different ISPs on its way to or from users. The reason is that in the Internet, most traffic passes through multiple ISPs on its way to or from the user. Even in an enterprise environment, traffic between businesses will traverse at least two different network infrastructures. To ensure consistent treatment of the traffic, the networks must be configured to treat the traffic appropriately on each network.

CHAPTER **2**

Service
Management
Systems

In the introduction we talked about some of the broad issues that tend to block the availability of network services. In the previous chapter, we described a way of looking at network services as a foundation for our examination of service management systems. In this chapter we examine considerations for service management software. The discussion begins with a working definition of what a service management system is and continues with a discussion of some of the functions that one should look for in a service management system. Later in this chapter, we investigate some of the factors that limit the availability of service management software. To some degree, the lack of this software inhibits the profitable delivery of services; as a result, we have a chicken-and-egg problem, and it is left as an exercise for the reader to decide which must come first.

What Is a Service Management System?

An effective service management system is one that fully integrates all the aspects of fault, configuration, accounting, performance, and security management across all the elements in the network that, either directly or indirectly, support the service. These service management systems can create integrated service level views, of the kind illustrated in Chapter One, that can transcend function, vendor, and technology.

Although many readers are familiar with an element or network management system, a service management system may be a new concept. These three systems—element, network, and service management—are in fact related and are built on top of one another. Specifically, a network manager can be built on top of an element manager, and a service manager sits on top of any combination of these lower level managers (Figure 2.1).

Element management is what many of us know best: we think in terms of individual network elements, typically from a single vendor, and of the software that controls their configuration and sometimes monitors their operation. We rarely think of element-to-element coordination and synchronization at this level, but many network operators are familiar with the problem of multivendor networks with a variety of vendor-specific element managers. To address the problem of uncoordinated element managers, some larger operators create their own man-

Figure 2.1
Layered
management
systems.

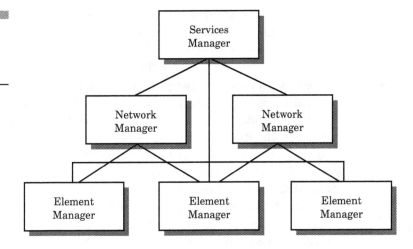

agement systems cobbled together from other systems. These are commonly built using PERL scripts.

Network managers are the systems we most often associate with general network status. These tend to be even more multivendor than are element managers and this wider range of vendor scope is due, at least in part, to the ubiquity of SNMP as a mechanism for fault reporting by managed elements. This reporting is often in the form of asynchronous messages sent from the managed devices to their assigned managers as SNMP TRAP or InformRequest PDU* [RFC1905]. Other SNMP protocol operations are also widely used as a method for collecting usage data.

---

*Asynchronous messages in the SNMP environment have evolved through the different SNMP versions. As a result of this evolution, there is often confusion about what is meant by a TRAP-TYPE or a NOTIFICATION-TYPE MACRO and an SNMPv2-Trap-PDU or InformRequest-PDU (Protocol Data Unit). Since this asynchronous capability is so important to building the types of service management software described in this book, it merits some additional clarification. In the initial version of the SNMP Framework, RFC 1157, A Simple Network Management Protocol described a Trap-PDU. As a general rule, it was assumed that *agents*, which we sometimes call *managed devices*, sent such asynchronous messages in a Trap-PDU to their managers. These messages were used to signal an unusual condition, such as an interface failure. An Informational RFC, 1215 [1215], A Convention for Defining Traps for use with the SNMP, gave guidance about how to define the asynchronous information that was to be conveyed in a Trap-PDU. It was defined with the TRAP-TYPE MACRO. TRAPS have remained an important part of the SNMP framework to this day. They are supported in SNMPv2c and the current SNMPv3. Since SNMPv2c [1901] introduced new data types, such Counter64, mapping to SNMPv1 is not always easy because it did not support these new data types. Although the wrapper around the basic message has changed, the definitions created with the TRAP-TYPE

Another important distinction between network management and element management is that network management systems are often aware of the relationships between the elements that make up a network; we say they are *topology aware*. They can be aware of topology at different levels, such as the network or physical layers. A common method for learning topology is to perform a *discovery* of the network elements through a wide variety of techniques. Some techniques actively probe the network and its elements, others read configuration data, and still others combine techniques. Some may actually provide a level of integration with element management systems. The end result is that these systems are able to report not only element failures, but frequently also the impact of a failure on other parts of the network environment. Such discovery engines are increasingly expanding their scope to include infrastructure services like routing or the DNS, but are not yet at the level where they can meet all requirements of a service management system.

Service management software casts a much wider net than either element or network management software. In addition to managing multiple aspects of the configuration and state of individual elements and the

---

Footnote continued from page 21.

MACRO formalized in RFC1215 can still be used to send SNMPv2c or SNMPv3 messages. One important characteristic of TRAP information is that the TRAP information that is sent using any of the versions of SNMP is not acknowledged. That is, the Trap-PDU or SNMPv2-Trap-PDU is sent and the system that sent the message has no direct way to know if that message was received. RFC 1905 [1905], Protocol Operations for Version 2 of the Simple Network Management Protocol introduced a several new PDUs. Of particular interest is the InformRequest-PDU. In many respects this is similar to the TRAP messages we have described. One important difference is that an InformRequest-PDU is acknowledged by the system that received it with a Response-PDU. This way the system that sent the message will get a confirmation that the receiving system got the message. As with all things, this is not perfect: messages, including Response Messages, can get lost, but in practice this is quite rare. What this gives us is a good mechanism for our service management system to use for the transmission and processing of asynchronous messages. Note that RFC1905 also describes an SNMPv2-Trap-PDU. This behaves just the way a v1 trap does but was needed due to the different wrapper around the basic message for v2. In this book, InformRequest-PDUs are assumed when sending asynchronous messages between systems. It is also generally assumed that SNMPv3 is used over the earlier versions since it has better security protection. An important point to remember is that starting with SNMPv2c, asynchronous messages were defined using a new macro, the NOTIFICATION-TYPE MACRO. This defines the information that will be sent in SNMPv2-Trap or InformRequest-PDUs in either SNMPv2c- or SNMPv3-based systems. Generally speaking, most systems provide local configuration controls so that users can select whether asynchronous messages go out using Trap or Inform messages. Since all of this detail is too cumbersome to include in the text when discussing asynchronous messages, we will use either a notification message or notification to indicate the general mechanism of asynchronous message definition and transmission in our system. Where a more precise meaning is intended, it will be described.

network they create, service management systems incorporate information about business issues such as contracts and service level agreements. (See Figure 2.2 for more details.) Note that a service management system can incorporate all the functions associated with element and network management systems into a single integrated system. Service management systems can also be created by gluing together many different element or network management systems. Gluing is in fact the most common approach in most networks, because there is, as yet, no "does-everything" piece of software.

The International Telecommunication Union [ITU] has done some excellent work in defining types of management. It has provided good descriptions of element and service level management as well as creating comprehensive descriptions of fault, configuration, accounting, performance, and security management. Unfortunately, unlike the IETF, the ITU's documentation is not generally free and is therefore less widely referenced in IP networking circles. Some relevant documents are listed in the reference section, along with a URL to the ITU's Web site.

Service Management Systems

Let's start the discussion of service management systems by describing what a service management system must do, and for whom it must provide these functions. As we saw in the previous chapter, a service can be a complex construct that is realized through the cooperative configuration and operation of many different kinds of devices and software, from Web servers to routers, to firewalls and much more.

As we can see in Figure 2.2, a service management system is responsible for all aspects of management. It receives input from a number of places, including order processing systems and customer and equipment databases. That data, combined with information about the network and its elements, is translated by the service management software into configuration parameters for each device and service.* Note that this is not a closed loop: information comes back from the managed devices about state and utilization so that accurate data can be output to service level, accounting, and billing systems. It is best not to think of this as a service

---

*This translation from high-level abstractions, such as business services, along with state and other information, into configuration information appropriate to each managed devices in the network is a critical function of a service management system. It has a significant impact on the system's design, as we will see in later chapters.

Figure 2.2
An integrated service
management system.

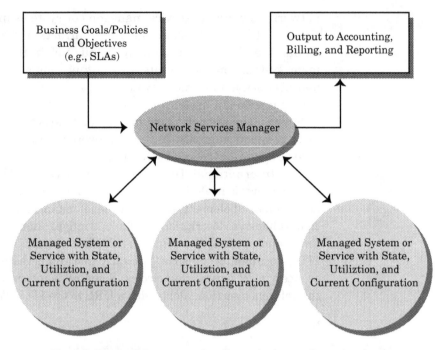

Network Services Manager sends policy and other configuration data
to managed systems and services. Manager collects state, utilization,
and verifies current configuration information.

management system and a bunch of devices. Instead, try thinking of it as
a series of management applications, databases, and network elements
that together constitute a system, because all of these components are
interdependent. (In later chapters, we look at this kind of distributed system in more detail.)

Service Management System
User Requirements

As we saw in the previous chapter, different organizations have different needs with regard to the service management system. In particular,
let's look at customers and service providers right now. In both cases, if
the basic needs of the service provider and the customer can be met, so
too can the needs of other parties involved in the service delivery and
consumption. Together, the customer and service provider have the

most demanding requirements. For example, the service provider must configure and keep track of those systems that provide the infrastructure services in addition to any other systems involved in the delivery of high-value services. It must also track accounting and state information so that accurate bills can be generated for the customer. The customer does not see all of this, but in our previous example, Acmeon was also concerned about the application server that Acmeon's customers had little direct visibility into.

What Do Customers Require from a Service Management System?

Very little. They require accurate bills and reporting as to whether they got the services they ordered or not. They are also interested in the stability of the service, which can be influenced by an effective service management system. Later, we take a closer look at some of the issues impeding effective service delivery and the software to manage these services. For now, here is a simple list of the customer-visible features that a service management system should provide:

* Accurate and timely billing information, without which accounting and billing cannot generate correct bills.
* Service verification—Did I get what I paid for?
* Audit information—Not only did I get what I paid for, but were the resources used in the way that I intended by the people for whom they were intended? For example, I might have purchased an additional leased line for a branch office to ensure reliable performance to the corporate order processing system for the sales group. If the majority of the traffic originates in another department at that branch and goes to a long list of Internet Web servers, the sales force might not have all the resources they need during critical periods. Even if Differentiated Services or some other approach is used to manage bandwidth, I will still want to know how much of each type of traffic I have configured the system to give priority treatment to. This is in order to make ongoing changes and verify correct operation. This is a good example of the important relationship of accounting and performance information to configuration.
* Responsive, reliable, service activation and change, so that the time between service order and service availability is minimized. Later, we

see how an effective service management system can help reduce the time to service activation and provide accurate data about when the service will be available. Many customers will not be aware of the details of this function, but only of how well it functions.

Some service providers have hit on the idea that giving their customers status views and online access to usage data is a selling feature. If customers desire it, and service providers offer it, then this is one more requirement for the system. (Along with this feature go a number of complex security issues, such as who can access the data and from which locations, and how the confidentiality of the data transmitted to and from the managed system will be preserved.).

Another extension of the aggregate service is allowing customers to reconfigure certain aspects of the service. They could, for example, be allowed to increase the amount of bandwidth or reserve network facilities, on a temporary basis, for special events—anything ranging from a series of high-level teleconferences to the bandwidth needed to accept orders after a nationally televised commercial. The conclusion to be drawn from this set of requirements is that it is desirable for networks to support dynamic reconfiguration of at least some components in the network as needs and demand change.

None of the features that allow customers to view data or perform reconfiguration operations, directly or via software that can perform such reconfigurations dynamically, can be responsibly deployed without a reasonable security infrastructure. The software must be able to ensure that people who wish to make changes are indeed who they say they are and that they have been authorized to make the configuration changes. RFC 2574 [RFC2574] gives a good summary of the types of threats to security that should be addressed in a management system and how SNMP provides mechanisms to guard against these threats.

What Do Service Providers Require from a Service Management System?

A service provider has significantly greater demands on the system, beginning with the fact that many departments within a service provider organization need to access information from it. Providing for such access can be one of the most challenging aspects of service management software, since the data often resides in different departments in uncoordinated databases. To continue with our Acmeon example from the first chapter, here is a list of departments that would potentially have interest in one or more aspects of a service management system, and the information and functions they would desire from it:

* **Physical infrastructure group**—Information about the physical topology of the WAN.
* **Routing or IP engineering group for the network**—This includes the interior routing configuration as well as the external routing configuration and the policies that go with it. Sometimes also included with this group are people with skills necessary for the management of the network just below the IP and above the physical layer.
* **DNS services**—Although there may not be a separate department dedicated to the management of name services in a network, there will usually be at least one person who is the DNS expert. This person needs much of the physical and routing information described, in addition to the location of the DNS servers and configuration information about each. In addition to configuration information, the people who manage the DNS will also be interested in the load on the DNS services, where the load is from, and the current state of the service (is it running correctly?)
* **Mail services**—The general type of information required by people responsible for mail services is quite similar to that for the DNS; for example, fault and configuration information. In addition, they need some of the DNS information to correctly configure the mail servers and identify faults when they occur.
* **Customer Web services group**—Like the DNS group, the Web services group will want information about the configuration, status, and utilization of the Web servers. In addition, they may also need information about the infrastructure, routing, and DNS services, because the Web services rely on these important basic services.
* **Colocation services for the application servers**—Some network operators have customer colocation facilities, in which they place a large amount of customer-owned or leased equipment running specialty applications. It is likely that people who manage these devices require tools specific to the application. For example, if the application servers incorporate database technology, some of the tools would necessarily be used for database management. In addition, some informa-

tion about the basic infrastructure services is also needed to correctly configure and manage these specialized services. Without this information, it is difficult to correctly manage these devices. If there is a failure in the service, it is necessary to determine if the cause is a network infrastructure failure or the result of a failure of the application.

- **Network operations departments that monitor the network and the services**—Often this department is subdivided as well. One group may manage the backbone network while another manages the Web servers and another manages dedicated application servers.

- **The group responsible for the data collection systems from which accounting and other usage and service-level information will be gathered**—This may not be the accounting group at all, but an operational group charged with polling data from the network for output to the accounting system. It is often the case that a network operations center (NOC) monitoring the state of the network is only concerned with the real-time operational status of the network. In some organizations, yet another group manages the systems that collect historic data.

- **The central engineering group**—An engineering function that is sometimes performed by the central engineering group is capacity planning or performance management. In other organizations, a separate capacity-planning group may perform this work. To make matters worse, those who undertake planning for the network infrastructure may not be involved in planning the resource requirements for servers. In this case, each department has its own machines collecting data, some of which is naturally duplicated. In the systems environment, the function of data collection and analysis to assess performance levels and predict future requirements is well developed, understood, and applied. When applied to network infrastructures, it is less well understood. To some degree, this is a result of the many dissimilar data-collection and analysis features used by network organizations. The lack of effective capacity-planning tools for the network environment also represents a cultural difference. Historically, data networks purchased more capacity any time they thought they would need it, as opposed to basing the purchase decisions on hard data that projected areas of difficulty. From an economic perspective, this approach no longer makes much sense, not only because of the direct cost of purchasing equipment that may not be needed, but because of the risk of failure to purchase equipment that is needed. Additionally, unnecessary equipment must still be configured and managed, thus further increasing operational costs.

In some service provider organizations, these groups may be collapsed or additional subdivisions may be needed. The list of operational functions is not complete; it considers just those groups that have a real interest in and responsibility for the operation of the network and the services that are deployed on it. We have left out the various management requirements and reporting functions; we have also left out the complex interrelationships between these departments and the data they manage. These data interrelationships are at the heart of the network and services management challenge, and they will be discussed later.

Now that we have listed the organizations that the service management software must support, we can begin to develop a reasonable list of the functions a service management system must provide. To facilitate this discussion, these functions are divided into Fault, Configuration, Accounting, Performance, and Security (FCAPS). This list is intended for expository purposes only, because an entire book could be devoted to the expansion of these FCAPS features. For those interested in more details, see the [M.3200] and [M.3400] references.* The following lists of questions in each of the main areas of management should be considered when planning what a service management system should do. They are presented from the perspective of the people who run the network.

Fault Functions

Fault-management software should help the network operator answer the following questions:

* Is each of the infrastructure elements functioning? If not, which are not?
* What is the impact of an infrastructure element failure on other elements of the network? (Information about the topology of the network is fundamental to any accurate diagnosis. Topology is also fundamental to almost all network management, so although it is referenced here, it is assumed in other sections as well.)
* What is the impact of any of the above failures on services, customers, service level agreements, peer network operators, etc.?
* What caused the failure (e.g., software or hardware failure, circuit loss, etc.)?

---

*The International Telecommunication Union [ITU] has written extensively on this subject. They publish documents in series, for example the M Series. A number of documents in this series are relevant to the present discussion.

- What is the estimated time to repair?
- Can backup paths through the network or other servers take the place of a failed component? This implies that the network has been provisioned with this capability in mind and that the service management software knows the difference between a primary path or server and one or more backup components.
- Are there any impending failures? Prognostication is a bit trickier, but some systems can predict a failure, based on known information, either before it affects a service or before real-time data confirms the failure event.

To support many of the functions in this list, the system must provide information about device configuration such as device model and other vendor-specific data. As we see later, configuration information is at the heart of any management activity.

Configuration Functions

Not only must the configuration software be able to correctly report the current configuration of a network and the elements it contains, but it must also be able to change the configuration of the elements in the network. Here's a list of configuration questions that operators need their management systems to answer:

- What is the configuration of each individual device in the network?
- How has the configuration of a device changed over time? Notice that there are several important reasons for including this feature:
  - Network difficulties often arise as a direct, secondary, or tertiary effect of a configuration change. By knowing when changes were made to the network configuration, operators can better track the source of network difficulties.
  - Configuration information is not only important to the fault management functions, it is also crucial to effective capacity planning. To know how much equipment or resources to purchase without over- or understocking, a network operator must know the network hardware and software and how configuration changes are likely to affect performance. A common example of this principle is a change in an interface card from a low to higher-speed version. If one did not, for example, record the time that the interface card changed from a 10-megabit-per-second to a 100-megabit-per-second version,

the utilization information would be off by an order of magnitude. *Utilization* is commonly calculated by dividing capacity by usage over a specific period of time. Without this information, the network operator may spend too much on unneeded equipment or too little on resources that will be needed.

– Billing and accounting present a third reason for recording configuration changes over time. Configuration changes are often associated with changes in billing rate, as when more bandwidth has to be provisioned for a service. Whether the order-processing system triggers some network event that changes network element configuration, or the change is accomplished via some indirect mechanism, all this information must be coordinated and fed back to the billing system before the account can be properly charged.

– Current configuration information is needed to help operators know if the customers are getting the level of service or performance they have ordered, at a particular moment in time. Historic information about configuration, along with utilization data, is used to show what the performance was an hour, day, or week ago.

▪ Is the device correctly configured for the service or services it is to support? One of the issues that makes services management more difficult than element or network management is that a service can't operate effectively unless many devices and underlying services are configured to work in concert. The Acmeon example service illustrates this point.

▪ Are element configuration changes coordinated across the network? Using coordinated configuration, the service management system can make a selected configuration change to many elements at the same time, thus causing a simultaneous change in a service (or services). For example, if a service provider offers VPN services for a customer and that customer wants to control the amount of bandwidth allocated for certain types of traffic, it may wish from time to time to change the amount of bandwidth reserved for one protocol or another. If the network elements don't get reconfigured in a coordinated way, one device may allow minimal traffic of a particular type while another device allows large amounts. Imagine a network that connects two offices in different cities: at each of these offices a device controls how much of each type of traffic (e.g., Web traffic to specific sites, email, and other traffic) is permitted to be carried over the link that connects the office to the network. If the devices are not reconfigured in a coordinated way, important traffic between the offices might be delayed at either office until the devices are synchronized to allow the desired traffic to flow correctly in each direction. This does not mean that each end must be configured identically, only that they need to be configured to allow the desired traffic to flow in the correct proportion in each direction. One device might allow a great deal of Web traffic to be outbound while the other might allow a lesser amount. If this coordinated reconfiguration is not done, the result will be undesirable for however long the network elements are out of sync.

Accounting Functions

A service management system does not necessarily have to incorporate the functions of an accounting or billing system. It must capture enough data so that accurate information can be extracted for use by external accounting and billing systems. That, after all, is the whole purpose of the exercise! Without generating accurate and timely bills to be paid by customers, providers can't accumulate revenue. At a minimum, the service management system should be able to report:

■ **Usage by service**—This gives planners and marketers information about what is being used by customers, and what is in demand.
■ **Usage by customer by service**—This is the aggregate by which some customers will be charged.
■ **Usage by customer by service by device or region**—Some service agreements may be structured on this basis.
■ **Time indexing for all of the above**—Service level agreements may be written to charge one rate during peak periods and another during nonpeak. Many operational support groups do their network backups in the early hours of the morning, on the assumption that the high demand on network resources that the backups entail will be less bothersome to users then. In the same way, a service-level agreement might offer lower rates for certain types of services during particular portions of the day, or particular days of the month.
■ **Failure or fault information**—Depending on the nature of the agreement, a single fault may not be significant. If the fault occurs a certain number of times during a specified period of time, financial or other penalties may be incurred. For accurate accounting at such times, the service management system must be able to record these failures precisely and tie them to both the customer and the services they affect. This task is even more complex than it appears, because the duration of failure must also be recorded. A hundred small "blips"

may be completely acceptable, whereas a single outage of an hour or more may have significant negative consequences for both the customer and the service provider.

* **Performance or latency information**—Like fault information, performance information is needed for every customer and service. If the provider has promised a certain responsiveness, the provider must be able to demonstrate that service delivery fell within these guidelines. One important difference between performance data collection and failure and fault data collection is the volume of the data; it is far greater in the performance realm.

Performance Functions

The type of data collected for performance is largely the same as that for accounting. The primary difference is that whereas accounting information is mostly associated with an individual customer and the services ordered, performance data is more general. It is collected to show how infrastructure services in general are functioning (such as the query and response times of the DNS), or to verify latency through portions of the network. These data are often used for capacity planning purposes. Whereas fault data is collected on an exception basis, as the faults occur, performance data must be recorded at regular intervals. The intervals should be just close enough for verification of the performance of the network infrastructure and the services that are built on it. Some systems collect data at a rate greater than required, which results in excess load on the network, the managed devices in the network, and the management system. Important questions that the performance portion of the service management system should answer include:

* **Total demand**—How much work of a particular type did the system or service perform? For example, what was the number of packets received on a specified interface, or how many Web pages were requested from a Web server?
* **Total utilization**—Utilization is similar to demand except it takes into account the capacity of the service, device, or network link. If we have a Web server that can deliver a maximum of 10 pages per second and we record that it served 300 pages over a 60-second interval, we would be able to say that the server had been 50 percent utilized. That is because, during a 60-second period, the server could have

delivered up to 600 pages, but in this case it had a demand of only 300 pages, or half of its capacity.

- **Trends and state**—Based on the utilization and demand figures, the system should be able to answer:
  - Are there any performance-related difficulties now?
  - Which systems and services are affected?
  - Where are future resource limits for systems and their subcomponents likely to be hit and how are these components related to other resources in the network? If one resource reaches its maximum capacity, how will that affect other resources in the network that are working together to deliver a service? The problem posed here is how to avoid increasing the capacity of one device to improve its ability to deliver a service, yet ensure that other devices also delivering the service will be able to keep up with the extra work that the more powerful device presents. Imagine again that we have the Web server that can deliver a maximum of 10 Web pages per second. If that Web server is connected to the network via a fixed-speed link and we increase the speed of the link by an order of magnitude, we may not see a corresponding increase in the number of pages served per second: the Web server can still only deliver 10 pages per second, even if the network connection can carry many more requests and potentially served pages.

Security Functions

The security dimension for a service management system has several aspects:

- Configuration and management of the access to operations on each managed device in the network. This is an area of significant complexity. For example, some organizations only allow operators in a particular group to make configuration changes. On occasion, nobody from the so-privileged group will be available; that means that another group, or specific members of that other group, may be granted extra privileges during certain periods. An operator not normally allowed to make a specific type of configuration change might be allowed to make such changes on holidays and weekends in some emergency conditions.
- Reporting and logging all configuration activities.

Notification and Reporting

Missing from the list above are the large number of features that display network information. These could range from dedicated user interfaces on workstations, to Web browsers, to textual reports printed once per month. We discuss these capabilities in later chapters on management software design.

The capabilities described here represent a daunting list. It is no wonder that no one place exists where a network operator can turn for a complete management solution—in spite of what we might have read in marketing brochures. Keep in mind that, at least for the present, such systems can be constructed out of many subcomponents, very much like the networks they manage. There are a number of benefits to this approach, most notably getting the "best of breed" from each vendor. There are also a number of significant problems with this approach, and service providers struggle with them every day. These problems include the complexity of supporting diverse databases of information and the difficulty of configuring the various systems that perform fault, configuration, and performance management.

Why We Don't Have Effective Service Management Software

Looking at the preceding section, you can see that, for service management to work, all the functions usually ascribed to system and network management have to be performed in an integrated fashion, they also must be tied to service subscribers, and the whole ball of wax then must be connected to the billing and accounting systems.

The functional complexity just described is a limiting factor to the availability of service management systems as we have defined them. The availability of complete solutions is limited even if we expand the definition of these systems to include those that have been cobbled together from a lot of different sources. This lack of available solutions, whether off-the-shelf or custom-created, in turn inhibits a service provider's ability to develop new types of services that could help with the all-important profit margin.

In the next section, we to take a closer look at some issues that get in the way of the development of service management software with the capabilities previously outlined. Only after understanding these difficulties will we be able to describe a more productive approach to creating service management software and provide some benchmarks by which such software can be measured.

The Utility Analogy

If we consider a telephone or electric company, we see similar requirements. For the most part, telephony environments have much more effective service management systems than do IP environments. As customers of the phone or electric company, few of us, even in large enterprises, have much insight into the inner workings of either. They have tended to be quite reliable over long periods of time, in part because that is what our expectations are of those services. The energy shortages in California have probably increased customer demand (at least for the larger customers) for better real-time reporting about the state of the electric grid and utilization trends. Shortages aside, here are some of the differences between those more stable services and a complex Internet service:

- Internet services have historically tended to be less stable than the offerings of the telephone or electric company.
- Internet services, even simple ones, are far more complex than services from other utilities.
- Part of this complexity is a result of interdependencies between the different services: without a stable routing infrastructure, it makes little difference if Web servers are running or not—people will not be able to connect reliably to the servers if the routing system is not functioning. A degraded routing system, although still technically functioning, can have significant impact on a customer's perception of performance.
- In the Internet, one service provider has a greater reliance on another than even telephone companies do in the post-deregulation period. The telephone network is more homogeneous than modern IP-based networks, and there are fewer companies in the space. Those in our telephone example tend to be stable, whereas Internet companies have exhibited considerable instability. This corporate instability is manifested in the services offered: when organizations are just starting out, they are not as proficient in managing services as they might be after some period of practice. In addition, organizations that are

under financial stress sometimes cut corners, which can affect the reliability of the services they offer. One direct way this happens is by the reduction of skilled personnel to maintain the network.

Factors That Impede Service Management Software Development

I am sure that some utilities experts would say that I've presented an oversimplified view of their systems. To some degree, this is probably true, because they too share in some of the difficulties network service providers face in the creation of new services and managing them in a way that produces profitable operation. In this section, we focus on issues that limit the availability of effective service management software.

Functions Spread Across Machines

In the previous chapter we outlined the different views that customers, and customers of customers, may have with respect to an imaginary high-valued service. In Figure 2.3 we have duplicated that infrastructure. Notice that the service provider may have several organizations and software components that must be combined to offer even a portion of the features described earlier.

The first problem we see in creating a service is that functions that must operate in concert to create a service are distributed across many machines. These machines must be configured to operate as a cooperating system in order for the customers to get what they expect. In the diagram we show configuration points alone, but a real-world system would also require the implementation of all the other aspects of management, such as performance, capacity planning, and fault management.

Vendor Differences in Functions

Operational and management differences between a Web server and a router are inevitable. The problem is that server platform vendors that offer similar capabilities also have significant product differences for

---

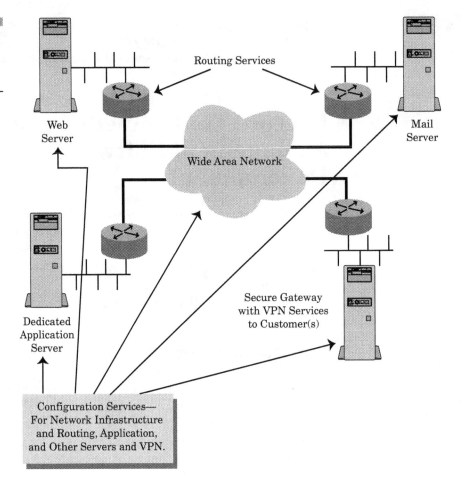

Figure 2.3
Coordinated configuration across machines.

similar functions. The same is true for router and other network equipment vendors. In some cases, these functional differences are the by-product of innovation and striving for customer responsiveness; in other cases differences are provided with the hope of providing some competitive advantage. In other cases these differences result from poorly written standards, specifications that may not be easily interpreted, or from "stretches" of the specifications for any one of a number of reasons. Regardless of the source of these differences, they make the management problem more difficult. Dealing with differences of all kinds is a key challenge to management software development and one of the issues that drive up the cost of development for management software.

Rapidly Moving Technology

Many of the functions now available or visible on the horizon were not available in commercial products just a few years ago. Whether these features prove to have lasting value is part of the problem for service providers and the people writing management software for them. Is the new technology worth investing management software development dollars in when there are no services that use them yet? This is a hard choice for management software vendors since, as commercial enterprises, they seek to maximize the return on their investment. For example, a number of technologies have been and are being developed to help ensure consistent performance of selected traffic through the network. History has shown that not all technologies are successful, and until the market makes one dominant, it is risky for an application developer to make a heavy investment in management software for that technology. This is an interesting conundrum, in that new technologies sometimes do not catch on as easily as they might because they are not integrated into existing management environments.

Even in those cases where the management software vendor is willing to take a risk, or has been persuaded by its customer base that a feature for managing a particular technology is important, technology itself is moving too fast for many vendors to implement meaningful functions.

A corollary to this problem exists with respect to the integration of the management of the new features with the management of older technologies from one or more vendors.

Variability of Access Methods

Later, we examine a number of different approaches for representing and transferring management data. What is relevant to our current discussion is that there has been, and continues to be, discussion about the *best way* to represent and transfer configuration information. For the most part, SNMP is well accepted as a data collection tool, but not for configuration. As a result, vendors have developed their own CLI and configuration file representations. A side effect of this practice is that new software releases, even from the same vendor, may have changes important enough to warrant change or replacement of the service management software.

Within service provider organizations, those responsible for router configuration have adopted scripting, frequently in [PERL], as a defense mechanism against this kind of flux. These scripts interact with a single managed device at a time and mimic the behavior of an operator sitting at a terminal typing in commands. The responses come back from the managed devices in a form that is intended for display on a screen for a human to read rather than as a program to read. These scripts perform *screen scraping*: they read the results intended for screen display and, based on those results, continue configuration activity or log an error. Operators can adjust the scripts manually by rewriting them, thus keeping step with the vendor as it releases new features. This is a reasonable survival skill, and so we shouldn't be surprised that it has become the method of choice, especially because there have been no other alternatives made available. We need to recognize, however, that this ad hoc manual rewrite of scripts lacks the rigor of a designed management system, has costs associated with these rewrites, and retains all the disadvantages inherent in an ASCII-based system—the most important of which is that ASCII-based interfaces, generally designed for interaction with people, make poor application programming interfaces for the development of sophisticated applications. To build intelligent automated management systems we need interfaces that are designed for machine-to-machine interactions and we should leave ASCII interfaces to their original purpose, which is human-to-machine interaction. These drawbacks have motivated some vendors to offer still other approaches that really do not solve the problem, but are refinements on the tools they have available.

Some developers are now looking to [XML] as the solution. As we see later, there is quite a lot to be gained from XML in some portions of a service management system. There are also major challenges with this approach, especially if you intend to integrate XML with SNMP and other management access methods.

Poorly Defined and Incomplete Standards

Before we try to understand what causes incomplete standards, we should establish what part standards play in the current scenario and provide a little background. Computer systems, especially very large ones, have been deployed in sufficient quantity and over a sufficiently long time to develop commercially available specialized tools and procedures for their effective management. We can cite examples of large networks, composed of complex systems, that are being managed top-to-bottom with sophisticated tools. Perhaps the best-known example of a successfully integrated environment is IBM's Systems Network Architecture (SNA). Another excellent example of network management in very large-scale systems is in the public telephone network, which has been extensively documented by Telcordia Technologies (formerly Bellcore). Of great advantage to these environments is that they tend to be closed systems with little need for management interoperability across technologies or coexistence between competitive management paradigms. This is clearly not the case in the IP networking world. IP networks grew up more recently than their large-scale computer system counterparts, and for these standards-based networks, integrated management systems are altogether a relatively new concept.*

The systems, protocols, and procedures for IP network management were developed without reference to most computer system management approaches. For an integrated service management system, this discontinuity creates the difficult problem of pulling together two separate disciplines.

Little or no integration exists within the historically separate domains of system and network management, in which we find waves of standards, standards bodies, and consortia of all kinds that have tended to fragment rather than unify the management environment. Some of these attempts have included the Open Software Foundation's Distributed Management Environment, the International Organization for Standardization's [ISO] portfolio of standards, the previously mentioned Bellcore standards, and Distributed Management Task Force [DMTF] standards.

The result of all this variety is that we have management systems fragmented in two dimensions. The first is the network and systems management dimension: there is one set of tools and technologies for managing our Web servers, another for key computing elements, and a third for our network infrastructure. The second dimension of fragmentation is the range of competing and overlapping protocols promoted for the first dimension. This diversity is not always the result of technical needs, but rather of the inability of different consortia to consolidate and coordinate efforts. In this somewhat chaotic state, it has been difficult to achieve what is both needed and desired: an integrated management environment, without which any service management system will produce disappointing results.

---

*It is true that SNMP has been on the scene for well over ten years but as we will see shortly, it takes more than just a management framework to enable the development of effective service management software. Many of these factors are nontechnical.

Over the years, the IETF has emerged as the preeminent standards body for IP-related technologies, yet even within this body, standards are often underspecified or missing. Whereas vendors that participate in the IETF can usually settle on a core set of functions for a standard, they always diverge on many of the details. The impact of disagreement on management standards—MIB Modules in particular—is that the IETF working groups attempting to define these standards find themselves restricted to just those management objects on which they can reach "rough consensus". Rough is often too rough, leaving huge gaps in the coverage of fault and statistical information within the standard. Vendors then create their own proprietary MIB Objects to fill in the holes, thus making each vendor different, not only in a functional way, but from a management perspective as well, and raising the cost of the development of useful management software once again.

Add to this problem the fact that standards for performing configuration management have been so contentious for so long that a new one comes along almost each year.* With each one, the already fragmented environment is fractured further. Some might argue that CLI is the standard for configuration, but there are two problems with this notion. The first is that there is no one CLI: even if a "standard" were created, it would suffer from the same underspecification problems we have seen in the development of MIB objects, which is often the result of nontechnical considerations, such as the misplaced belief of vendors that they should differentiate themselves by making their management instrumentation unique. A better place for such differentiation would be in the development of management applications. The second problem with the idea of a CLI as a standard for configuration is that, after years of this approach, we still have a poor suite of management applications. This is in part due to the inappropriateness of an ASCII interface as a programming interface, as mentioned earlier.

Instrumentation Differences between Vendors

Different standards and basic functional differences between vendors lead inevitably to differences in the management "knobs and dials" each vendor provides. These knobs and dials can vary even for the same fea-

---

*See [COPS], [COPS-PR], and [DMTF] as examples in Appendix A.

ture. Operators are resigned to the fact that to achieve a certain operational result, they must configure one vendor's product (router, Web server, or whatever) differently from another's. These mounting differences, despite general recognition of how unproductive they can be, are so pervasive that they can be found even between software releases from a single vendor, or between two models provided by the same vendor.

The View of Management Software as an Expense

When those who make purchase decisions for network equipment are asked, they often provide the following list of features as paramount in the decision process:

- How much does it cost? What is the cost per bit moved?
- What is the port density? (Footprint is a key concern.)
- How much power and cooling does it use?
- Do routing functions (to use a router example) interoperate with vendor A and vendor B?

Seldom, if ever, does the discussion move from here into the management domain. When it does, it is often perfunctory—a check to find out if the vendor has a CLI and some basic MIB objects. It is clear that these two basic capabilities, however important, do not provide even the foundation of a useful service management system. In that same vein, it is interesting to note that those primarily responsible for network equipment purchasing in larger networks are generally not the same people who run the NOCs or the accounting and billing departments. They don't create new services or market them to new customers; they are, as they have been since the beginning of IP time, the people who configure and deploy routers. It is reasonable that they would take a view that is somewhat focused and uninformed by the concerns of other departments.

Vendors cannot be blamed for slighting management software as long as they're reflecting the relative importance placed on the management function by decision makers within the organizations of their largest customers.

Management Software Funding Models—The Shortest Time to Revenue

The developers of management software of all types are subject to the same time pressures as their counterparts who develop the functional aspects of network products, from Web and database servers to routers and switches. They feel real and perceived pressure to develop a product as quickly as possible and begin to generate revenue, even if it means sacrificing future development efforts and even more revenue as a result of poor early engineering attempts.* Unrealistic time frames are a given in software development, and they're not all bad. But considering all the other drags on software for service management, time pressure must be regarded as detrimental, and it must be managed for downside risks. Under these conditions, it is difficult to create effective, scalable management software of almost any type, much less an integrated service management system.

Organizational Issues

In this section, we examine some of the organizational issues that inhibit the creation and management of profitable services in provider organizations. Although not every organization is afflicted with all of these concerns, nor are these issues a function of networking and network technologies, they still exist to some degree in many organizations of all kinds.

Intraorganizational Issues

Most network organizations are structured by function. Certain groups manage the servers, the routers and the core infrastructure. These

---

*Commercial entities are not the only organizations that suffer from a rush to release product. In the IETF, there is often pressure to have short time frames for working group activity and release "product" as soon as possible. In the case of the IETF, this product is a standard. There are many good reasons for this time pressure, such as causing a focus on a specific problem rather than a wide-ranging unfocused technological exportation by the working group. Whatever the reasons, time pressure does affect the output of working groups.

groups are often subdivided, meaning that those who configure routers in large networks are often not the people who take shifts in the NOCs.

A new trend has emerged to make this condition worse: some organizations are creating new groups to configure the more complex devices at the edges of networks. Their focus will be squarely on new services like bandwidth management and VPNs.

Although all of these groups often belong to the same corporate entity, they have different priorities, skill sets, and cultures. The result is that communication among them is not always effective, and the systems they build tend to be poorly integrated. There is no technical solution to this problem except to offer software that is highly modular. Modularity often raises the cost of development and deployment, because it takes longer to develop and requires more hardware and software support.

Accountability across administrative domains. High-value services often must be carried across multiple service providers, thus the relationships between providers of service must be clearly identified, with close attention to what happens when a fault occurs and how rapidly it is to be repaired. As aggregated services are deployed, this issue of ill-defined relationships will become more complex. Suppose a peer* of the primary service provider has a failure that causes an outage for a customer. If the peer's outage does not of itself drop the customer below a specified service level, that is one thing; but if that outage, in combination with other outages that are within the control of the primary service provider, drops the customer below a specified service level, the situation becomes less clear. Should the penalty to be paid to the customer in cash or reduced fees by the primary service provider, or should it be split with the secondary service provider that caused the service to drop below promised levels? The problem presented here is overly simple since, in the real world, more than two service providers may be involved.

Why do we need new services and service management systems? In the face of all the problems just outlined, there must be compelling reasons to attempt the creation of new services and related service management systems. *The only sustainable justification is increased revenue.*

---

*The term *peer* has several meanings. In this context, it is intended to mean another service provider that shares Internet routing information with the primary service provider. It is likely in this case that they will share routing information via the Border Gateway Protocol [RFC1771] that includes a specific meaning for the term *peer*. It is also assumed in the above example that these two service providers carry traffic for each other.

If we look back to the functions and requirements described earlier, we see that a service offering is the result of establishing a series of complex relationships that must be managed in a coordinated and timely way:

- **Relationships between different parts of a network device—** Access control may be configured in one part of a device and the application in another. To produce the desired result, both should be configured as an integral unit, because a possibility of error exists to the extent that they are not. You may be surprised to learn that many current systems may not offer a single access point to this configuration information, so the management system must coordinate separate parts to facilitate coordinated access control.

- **Relationships between different network components—**In some respects, this is the traditional network view as opposed to the device-level view that concerns itself with the operation of a single device at a time. In this network-level view, we are concerned not only about the relationships of network components, but also with those relationships in the context of a particular service.

- **Relationships between different services—**If the DNS is not functioning, mail messages to customers may be delayed. Relationships between services with the potential to disrupt more than one system can occur between any combination of the infrastructure or high-level services described in Chapter 1.

- **Relationships between the service provider and the customer—**These relationships are generally specified in the management application and can range from the details of services and service-level agreements to past billing information.

- **Relationships between service providers—**Many service provider networks may be traversed on the way to delivering a service to a customer.* For peer relationships to work, service providers need to make agreements about the amount and types of traffic that they will carry under a variety of circumstances. The terms of these agreements must be available to operational staffs as they troubleshoot problems and plan for growth.

- **Relationships between customers and their customers—**In an earlier example we looked at some relationships between customers and

---

*Even in the case of an enterprise with a private network, this problem exists. As we have pointed out, the backbone may be run by one organization, and different departments may run other pieces of the network. In this case, each department might be thought of as a different service provider.

their customers. These relationships can include from which networks the customers of the service provider customer will access the service to service-level commitments about latency. This type of information is needed for everything from billing to correct provisioning of the VPN.

- **Relationships between service provider organizations**—For services to be correctly deployed, managed, and billed, different organizations within the provider must cooperate. Historically, they have all had their own systems and network management approaches, which were poorly coordinated. These relationships may be further complicated by security policies at each provider that specify by whom, and under what circumstances, certain types of customer information may be viewed.
- **Data relationships**—We will eventually investigate the relationships between fault, configuration, performance, and other types of data in some detail. For now, one important example is the association of usage information with configuration changes and time of day. To produce accurate bills on schedule, the biller needs to know what work was performed on behalf of a customer, at specific times, for each piece of equipment involved.

Data Relationships

As it turns out, despite all the fuss about management protocols in the IETF and other organizations, effective management software remains elusive. One reason is that the working group focus, which should be directed toward these data relationships and the database technology to hold them, has too often been placed elsewhere. Work is now taking place in the IETF and DMTF to model a small portion of data relationships.

Data relationships enable key functions in an effective service management software system, and we'll describe those functions next.

Effective service management requires coordinated configuration.   For the service to operate effectively and be "turned on and off" at the right times, configuration activities must be coordinated across many systems of different types. The greater the time lag between configuration operations, the greater the opportunity for one portion of the system to get out of sync with another. The problem of coordinated configuration can be more or less difficult depending on the number and type of configuration systems entailed in configuring the various devices (and services they contain) that are used to realize a high-value service.

The more differences between the technologies and vendors, the more configuration tools will be required.

Coordinated configuration should not operate in a vacuum. Making good configuration decisions requires a good deal of information about other aspects of the system and a great deal of historic data:

- **Current operational state information** about the elements and services to be configured: it may not make sense to turn on a service if one or more of the elements or services required to realize the high-valued service is not available.
- **Current provisioned state** of the devices and services: if a resource is already spoken for, it may be unable to provide the services stipulated in the new configuration request without compromising existing service.
- **The duty cycle** of the devices and services: increasingly, time is an important configuration parameter. The configuration software must not only convey the desired times of operation for the new configuration parameters, it should also consider the possibility of conflicting duty requirements for services already configured in the device and network. This timing element applies to all configuration operations. For example, if a device is configured with an element in a down condition, that is fine if the configuration operation takes place during a maintenance window. In other cases, the fact that the component is down will be much more significant.

In addition to all of the above points, the following information can be crucial to the ability of a service management system to correctly manage existing and new services:

- **The capacity**—Knowing about the capacity of a device or service helps avoid the problem of oversubscription. Because performance often doesn't decline in a linear fashion, the management software ideally can learn some of the performance characteristics of each device.* If the management software is aware of a device's used capacity and knows something about its performance characteristics, more effective use of existing network resources can be made without the cost associated with overprovisioning. Those costs are twofold: the

---

*The most useful way this learning will take place is through observed performance of the network elements in normal operation. Some management systems might allow the input of thresholds in a manual fashion, which is a start but not nearly as useful as observing the devices in use under load at specific times of the day.

obvious one is the purchase of unnecessary equipment, but the greater cost is the ongoing maintenance and support of extra capacity. A well-designed and deployed service management system can help reduce both of these costs.

- **The capabilities**—It makes no sense to attempt to configure a device for a specific function if it is unable to perform that function. This is not as silly as it may sound: as network devices become more complex, their variability from one vendor to the next, and even one model to the next, increases.

- **Current utilization**—We need to know about current utilization for the same reason that we consider state information. If a device is near capacity, you may not wish to add to its current load. This issue is somewhat more complex when you consider that relative utilization varies from hour to hour and day to day. These factors should also be considered before the configuration of additional services on a device.

- **Roles**—In some cases, we may need to associate only certain parts of a device or service with a customer, while subparts of a device or service are reserved for other customers. The configuration software must be aware of these "reservations" and act accordingly.

- **Historic utilization and trends**—Part of planning a service offering is predicting whether a new service, or a change to an existing service, will hasten the exhaustion of a network resource. Prediction is only possible, though, where usage trends for the resource are known. Under most circumstances, it is good practice to set up a service on a device with the capacity to support the service for some time. If the device falls short of requisite capacity, additional configuration changes will be needed either to move the service to another device, or to add additional resources to the existing device. In either case, the risk of service interruption is increased.

Effective service management requires coordinated data collection and accounting. A basic objective for businesses is to be as profitable as possible. For an enterprise, knowing what departments use particular resources can help profitability planning. New resources can be added or deleted, depending on the business justification. For an ISP, the inability to account for usage by customers can cause significant underbilling.

Coordinated data collection and accounting is an essential component of any service management system. Whereas most network management products collect usage data based on interface counters and other fairly simple information structures, service management software must be smarter. It requires data collected on the basis of the configuration information for each device and service. Let's go back to the Acmeon Web service example we have been using. The service may have been created with a finite amount of disk space. It may also have been created with a service guarantee for a fixed number of hits per unit of time. To correctly bill the customer for these services, data must be collected from exactly the right Web server, running on a device that could very well be supporting many Web servers at once. Information must also be recorded about the average page hits per unit time to determine whether or not a surcharge is in order. (This depends on how the service level agreement is written.)

Once all data have been collected, they are output to accounting and billing systems that determine how much to charge or credit, based on the data provided for each customer.

Effective service management requires coordinated fault and root cause analysis. Many of the reasons for coordinated fault and root cause analysis are the same as for data collection and polling. When we take a service-level view of a fault, the root cause may be a component of the network that is far removed from the customer. Only by building into the system the relationships that allow for effective root-cause analysis at a service level can we hope to get effective and timely information about failures.

For the purpose of this discussion, the distinction that makes a root-cause analysis system (as opposed to one that just reports faults such as interface failures) is in the amount of value-added analysis and information provided. Many systems can report interface failures with a reasonable degree of reliability because they are based on the receipt of a single SNMP InformRequest-PDU or Trap-PDU [RFC2571 and RFC2572]. They provide very little "value-added" analysis of the data.

In a true root-cause analysis system, alerts may come from several devices in the network, which indicate an inability to communicate with a failed interface. In this type of system, information from the system that has the failed interface, along with the data from all the systems that can no longer reach this failed interface, is analyzed. A root cause of the problem—namely, the interface failure—is reported on the basis of that analysis. The advantage this type of approach has over simple fault reporting is that it saves valuable operator time and reduces the mean time to discover and repair the failure. Systems of this type can filter out the second-order problems and report the interface failure in such a way that the operator's attention is focused on the failed element.

Now let's apply this technique to an integrated service-level management system. Such a system will have all of the features just described (coordinated configuration, data collection and accounting, and root-cause analysis). In addition, relationships that create dependencies among certain services are well understood. In the Acmeon example, part of the service the customer purchased is email confirmation of orders placed on the Web server and validated by an application server that has inventory and other vital information. Should there be a failure in the DNS server on which the mail server relies, we would want a root-cause analysis system to "know" these relationships and report them in a way that is meaningful to the network operators. Some of these relevant relationships are:

- **Customer relationships to a service**—With this information, the root-cause system can let operators know when a service has been affected and notify customers within agreed time limits. It is easy to imagine a system that could automatically notify customers of such failures.
- **Service-to-service relationships**—As we have seen, a service, say an email service, might depend on one or more other services. In our example, the mail service relies on the DNS server. What we have not mentioned to this point is the obvious role that topology plays in this example: if DNS requests are not being met, it may not be because of a failure of the DNS server itself, but because of an infrastructure failure. One example is a cut fiber link in the path between the mail server and the DNS servers used by the mail server. Of course, redundant paths and name servers could help, but even so, we would want to know about the fiber cut and its potential effect on the customers and services that might use it.
- **Device-to-service relationships**—Devices have relationships to services as well. A single device might be used by many services. These could be different types of services or multiple instances of one service. A common example is the server platform that runs many Web sites for different customer companies.
- **Device-to-customer relationships**—The device-to-customer relationship extends the device-to-service relationship. If the Web server fails, we need a way to let customers know about it.

Scarcity and Cost of Trained Personnel

Management software can improve the effectiveness of NOC personnel in ways that have two positive financial impacts. First, the NOC may need fewer operators to support the same-size network since they will be less distracted by secondary and tertiary failure information and can get right to the root cause of a problem, which will generally clear most of the second order failure reports. Second, as more service level agreements are made, operators are more often obliged to report failures within a certain period. They may also have to guarantee a certain percentage of uptime. Failure to meet those guarantees can have significant financial consequences, to say nothing about the blow to reputation.

Nonetheless, if there were a ready supply of personnel able to develop and deploy the sophisticated managed services that we have just explored, the need for sophisticated service management software would diminish only slightly. When one multiplies the per-person cost of such staffing by the number of staff needed to deploy enough services to be profitable, cost eats into profit at an alarming rate. This is especially true if staffing costs are compared to the cost of software to improve productivity and reduce the need for these personnel. The larger the network, the more significant the savings.

To the extent that our scarce and highly skilled personnel can be freed from some of the more mundane aspects of service management, they can be deployed to more interesting and productive work—such as developing new services for sale—to further improve the profit margins of the network operator.

CHAPTER 3

The Edge and Service Management

When people talk about "the edge," they're usually thinking about the interface between a service provider and a customer. In fact, there are several different meanings for *edge* that can help illuminate the challenges of delivering high-value services in today's networks. In this chapter we expand on this discussion of the edge and its role in service management systems.

What Is the Edge?

According to the most common view, the edge is where the lower-speed access lines get aggregated by a device into higher-speed lines at a service provider location or in an enterprise. These higher-speed lines generally move away from the edge and point to the center, or core, of the network. For our purposes, we'll expand this notion in two dimensions. First, *the edge is not a single interface between a line and a device* such as a router. The edge is often a collection of devices and the media that interconnect them. These devices are often of different types, such as routers and bandwidth managers or firewalls that provide security. Regardless of the number and types of devices, they must be configured as an integrated system to perform any one of a number of value-added services such as bandwidth management or secure access to Web servers. Second, edges are often defined administratively and so can exist almost anywhere in the network. What follows are three common examples.

A Low- to High-Speed Aggregation Point

How can you, as a manager, guarantee a specific performance level as a service moves from slower links at the edge to the high-speed links more commonly found toward the core? Higher-value services are likely to have service level agreements with constraints on performance variability and guaranteed minimum throughput levels. As Figure 3.1 illustrates, slow(er) speed lines get aggregated together at the edge. The delay characteristics between the two technologies for the low- and high-speed interfaces can be quite different. How these links behave as they near their limit is also likely to be different due to these technology differences.

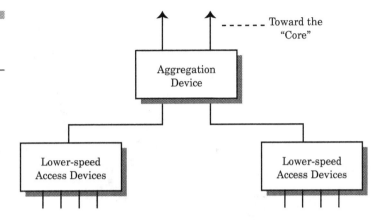

Figure 3.1
Aggregation of low-to high-speed links.

The important point is that *services traverse networks with different speeds and technologies.* Slower-speed interfaces get aggregated at one or more levels. The services, interfaces, and devices over which the services travel at each level of the network must be properly deployed. They must then be configured and monitored to guard against the differences in media and the performance characteristics of aggregation devices to ensure correct service performance. This point is one of the challenges facing an effective service management system. Routers can be made aware of the topology between devices and how that topology should affect the router code's decisions about where to forward the packets. In cases where a service is layered on top of a network infrastructure, the configuration software must consider the topology as well. Don't plan to do this step manually: the configuration software should at least assist in the configuration of these services, because the details are so numerous that very few humans could keep them in their heads.

Administrative Boundaries

Another important type of edge exists at administrative boundaries between organizations or departments. In the previous chapter, we discussed some of the organizational issues that crop up when a service is deployed across administrative domains. The security issues related to service delivery across administrative boundaries will be discussed later in this section. For now, we are concerned with coordinated configuration, data collection and accounting information.

Figure 3.2
Administrative
boundaries within
an organization.

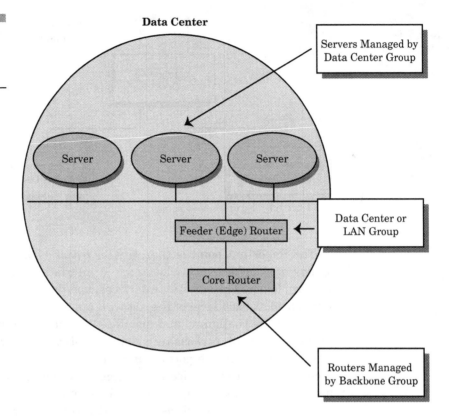

Figure 3.2
Administrative
boundaries within
an organization.

In Figure 3.2, we identify different organizations that configure and control the edge versus the core routers. Notice that in this case, the edge is a router that exists in the data center, probably inside an enterprise, and not at the boundary between a service provider and its customer, as one might have inferred from the previous diagram. Not shown, but also quite common in networks, is the administrative edge between hosting or colocation facilities and the rest of the network. A diagram of that topology would look very similar to the Data Center Diagram in Figure 3.2. Enterprises have long been familiar with the operational issues for this type of edge. The staff who manage the servers and computer platforms are seldom the same people who control network resources. A successful service management system is one that is successful at bridging these administrative boundaries. This bridging function is essential if the customer is to receive the desired service. The example in Figure 3.2 shows a single *core* router in the data center, controlled by the backbone group. Some organizations use one or more *feed-*

er routers in this type of environment (what one might find at a data center or some other edge point, as shown in the diagram). These feeder devices are often managed by a group separate from the group responsible for the core devices. Therefore, in Figure 3.2, there are three rather than two organizations configuring this portion of the network (namely, the core router group, the edge or LAN group, and the server group). And even where there are still only two, the server or data center personnel may be responsible for the access devices, while the core group retains responsibility for the core devices.

You may recall from the discussion of organizational issues in the last chapter that there is yet another axis to the problem of administrative boundaries in an organization. Competing organizations aside, the organization that configures a router usually has a suborganization responsible for the day-to-day fault management and collection of data from the routers.

A customer's network traffic may have to traverse several service providers to get to the application servers supporting the particular service purchased. In most cases, the customer won't know anything about relationships between service providers, and most would prefer not to have to know, as long as the network works. Service providers have had *peering agreements* for some time. These agreements mostly handle issues associated with the exchange of routing information, but for effective service delivery, they must be considerably more intricate. Each provider has to monitor the services that flow across its administrative boundaries for its own customers (see Figure 3.3). This monitoring is necessary to guarantee that the services operate with the performance, security, and reliability promised to that customer at the time of purchase. But if the customer is not "theirs," providers may also need to keep track of the traffic that flows through their network as a result of expanded service agreements. This increased monitoring is necessary to ensure that they meet their obligations to their network peers. It is important to note that it is presently not practical for service providers to monitor each customer flow they carry for their peers—but they must monitor the overall service quality of different types of traffic across administrative edges. Similarly, service providers are required to keep customer-based data on their own customers, not the customers of their peer service provider. To accomplish this higher level of cross-provider service verification, providers will have to expand the types of monitoring that they have historically performed.

Figure 3.3
Boundaries—cross-
organizational
management.

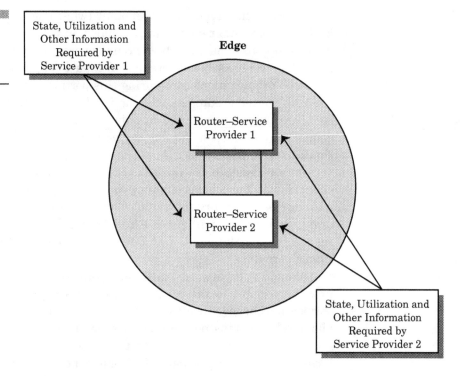

Figure 3.3
Boundaries—cross-organizational management.

The Interface between Server Environments and the Network

Now let's talk about the very important edge that exists at the interface of the network to the server environment. Figure 3.2 introduced some of the organizations that may be found in this environment. Now, we extend that discussion to include products that might be used to deliver the base services on which the more complex services can be built. This edge not only includes the transition of administrative domains (WAN to server groups, for example), but also includes technologies that are not part of a core routing environment at all (Web load balancing, for example). These newer technologies are being used more and more often as the key services on which high-value services can be built. Similar technologies might also be involved in peer-to-peer edge scenarios, and we'll get to those in a bit more detail in the next section. For now, note that in Figure 3.4 we see differences in technology as well as in the groups that manage them.

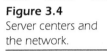

Figure 3.4
Server centers and
the network.

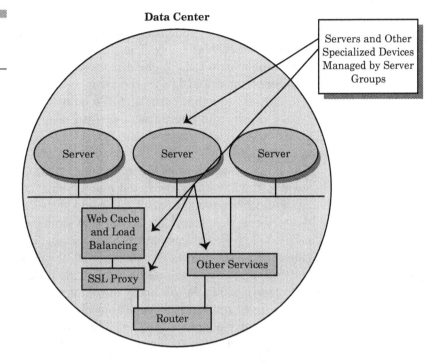

Complexity and Technology Convergence

Figure 3.4 is a good introduction to the technical complexity that is
steadily developing in service networks. Disparate technologies residing
at network edges present management systems with a problem; there
has been a tendency to push new technologies to the edges of networks,
where speeds are often slower. (New technologies are sometimes introduced in lower-speed versions, and their performance increases as the
vendors learn and improve the technology.) An important reason for
increased technical complexity at the edge of networks is that the people
who manage the core want to keep it as simple as possible—by pushing
newer technologies to the edge. This has proved to be a good thing
because it also makes sense from a topologic perspective: if an
[SSL/TLS] proxy device is used to improve the performance of secure
transactions to a web server, it's desirable to place that device near the
Web server. This approach makes for a simpler core, which can focus on
the work of moving packets around rapidly and efficiently. Router configurations are not complicated with the additional work of configuring
QoS, Web caching, or VPNs.

There's bad news too. At the edges, which can be any of the edge types we have defined thus far, enormous complexity makes service creation and management hairy. There can be new technologies needed to deliver a single aspect of a high-valued service, such as QoS for example. Both Differentiated and Integrated Services might be found at the edge, used to help ensure that certain traffic transiting the network gets the resources required to guarantee a certain level of responsiveness. The edges are also logical places to configure VPNs and perform a range of proxy operations such as SSL/TLS.

Figure 3.4 shows some of the technologies that are likely to exist at the edge and will need coordinated management if a complex service is to function correctly. As it turns out, many systems are dedicated to one or two services. For example, we find that routers do not generally perform SSL proxy operations, and the SSL proxy devices do not as a rule provide routing functions. The problem this creates at the edge is that not only do network managers have to juggle different technologies, but they also have to juggle the many different systems that are delivering each of these technologies. In addition, the topology used to deploy such systems can have a significant impact on how the network behaves. Service management software must be aware of those topologies so that it can report failures correctly and collect adequate information for long-term capacity planning and performance management and avoid future bottlenecks.

It stands to reason that if you accommodate a variety of systems that must be configured and managed at the edge, you will end up working with different vendors, each with different configuration and monitoring parameters. This may be true even when the service is one that has been standardized, because the standards are almost always expressed in terms of on-the-wire behavior and not in terms of the configuration dials. And keep in mind that even when standards do exist, as we saw in Chapter 2, they are often incomplete.

The disparity between relatively low-speed and higher-speed interfaces is another problem specific to the edge. Once again, service management software must know the relevant differences, because different media behave differently depending on their load.

Security Issues at the Edge

Of all the issues the edge presents, one of the hardest is security, because as we begin to develop higher-value services, the intrinsic value of the data these services carry increases. At the same time, people who use and manage these services will need access to a lot of potentially sensitive information about the service, how it has performed, and how it is performing. In some cases, they will need access to the data of the service itself, for verification and other purposes. Figure 3.5 illustrates some of the types of access required.

Figure 3.5
Intra-organizational
security.

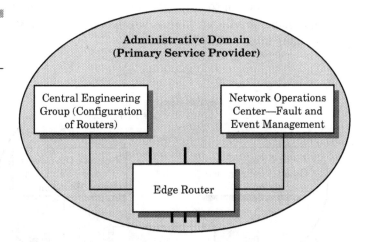

As we can see in Figure 3.5, we can encounter significant security issues even within a single administrative domain. There are two aspects to this problem: one for those administrative groups that generally have access to a device such as a core router or server, and one based on the function or role of the individual.

In our example, the central engineering group is responsible for all router configuration activity. As is the case in many organizations, they also are in the escalation chain for problem resolution, meaning that when the NOC has difficulty resolving a problem, the central engineering group is called in to assist. For this reason, the central engineering personnel need read and write access to all information in the router, including potentially sensitive usage-based counter data.

The box at the top right of Figure 3.5 represents the front-line network operators. Their responsibility is to monitor the daily operation of the network and report and repair faults as they occur. Most of these operators, at least in larger networks, do not perform the day-to-day configuration and planning of routers and other network devices, and therefore they don't have the same write access to configuration parameters as do the central engineering staffs. Unfortunately, circumstances are rarely this straightforward; in some cases, senior personnel in the NOC, or available to NOC staff, have emergency configuration authorization. Ideally, the security systems in place have the intelligence to know who can perform what activities. In some network environments, privileges change with the time of day, as when the control moves from NOC to NOC around the world, for example, from San Francisco to Boston to London to Sydney. During the time Boston has control, it may enjoy more privileges than it will when control passes to San Francisco as it gets later in the day.

Figure 3.6
Security across administrative domains.

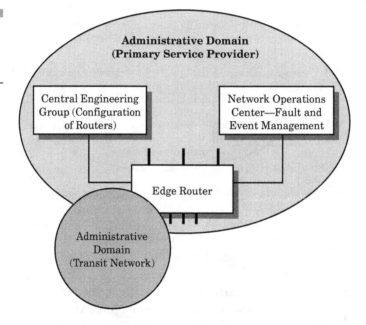

Security concerns increase when an edge device sits as the gateway between two administrative domains, as shown in Figure 3.6. Here, an edge router has interfaces that connect to edge devices belonging to a peer ISP that provides transit services for the primary service provider.* It is also likely to be true that the primary service provider reciprocates by affording similar service to the secondary provider's traffic. Although the primary owner of the device may not allow anyone else to make con-

---

*We could have a very similar scenario between two enterprises with connected networks that do not go through the Internet. This is not uncommon in the absence of good service-level guarantees and security services in the Internet.

figuration changes on the device, it may have to allow monitoring activity for at least those interfaces connected to peers. Beyond basic monitoring, the primary service provider must allow enough data collection to verify certain contracted service levels by the secondary service provider.

Figure 3.7
Customer access.

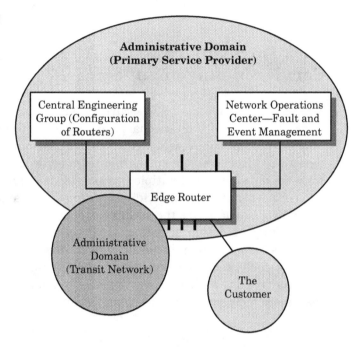

The legitimate needs of the customer represent as much of a problem for security as the other examples we have discussed. Customers expect access to information about the services they have purchased and, in some cases, the devices that deliver those services. Often, they also desire aggregate information available to them online about current and past usage. In Figure 3.7 the customer directly accesses information in the router as opposed to passing through the transit network; thus, customers may have access to some information not available to personnel from the transit network. It is, however, likely that some of their traffic will pass through the indicated transit network.

Another dimension to consider in the security domain is that not all of the customer's personnel are necessarily granted the same access rights. Different users accessing the same device may be restricted to different operations and sets of information, just as operators within an administrative domain may be. In the case of an enterprise that runs its own network, the central network administration people might set up and configure routers at branch offices, but the local area network (LAN) personnel may also want access to at least some of the information in the edge device. In this case, the customer is the branch office and the service provider is the enterprise's central engineering staff.

How Do Security Issues Affect a Service Management System?

Throughout this discussion of security we have focused on the organizations and those of their subparts having a legitimate need to access information on edge devices. A service management system, whether constructed from a single monolithic set of software or comprising many software modules from multiple vendors, must be able to distinguish the different organizations, individuals within them, and the roles of each individual, to guarantee the security of sensitive information. As we get deeper into the design of management software that runs in managed elements like servers and routers, we'll discuss how to allow the integration of security facilities for services at the edge. As we address applications for creating a services management system, we will also delve into their more complex security issues.

Why the Edge Is so Important to a Service Management System

By now it should be getting easier to see why the edge is pivotal for service creation and management. The edge is the nexus of technologies, service organizations, and customers. These combinations take on a multitude of forms, each presenting the administrator with a different challenge. A service management system must be able to handle each combination in a scalable, efficient, and secure fashion as it arises, to assure the correct configuration and operation of customer services. Moreover, it must collect data to record the units of work performed by the services for later output to accounting and billing systems.

Nobody likes complexity in his network, and high-value services introduce complexity that cannot be eliminated. Successful service management systems will be those that are smart enough to understand the complexity and help make appropriate management decisions in a way that offsets the cost of the complexity, to a degree that it makes it profitable to offer the high-value services. To the extent that we are unable to develop service management software that can help in this regard, we are faced with a strong disincentive to create new services in our data networks. This reluctance to develop new services takes its toll on the profitability of the organization that runs the network.

Policy and Service-level Abstractions

To meet the challenges described in Part 1, we must change the way we've been thinking about the network, the services we wish to provide on that network, and the management of those services. We open Part 2 by corralling the core concepts of policy-based management, which offers some techniques and technologies that can be applied to the problems previously described. Following that discussion, we'll propose an object-oriented model for the creation of a service delivery system. This sequence lets us work on the problem without getting tangled up in the technologies that might be brought to bear on solving it. Too many good discussions about the general nature of service management software requirements dissolve prematurely into debates espousing one technology over another. I expect some readers to have strong views about the best technologies for one aspect of the management solution or another; they can use the requirements developed in this chapter to measure those favorites and see how well they stack up.

CHAPTER 4

Policy-based Management

A great deal has been written about policy management in recent years, and many believe that it can be an effective element of service management software. This chapter* is a brief overview of policy-based management as it intersects with high-margin services and the software that controls them. For readers requiring a more in-depth discussion of policy-related concepts and technologies, the references section contains further pointers.

What Is It?

Sitting atop the Tower of Babel of networking nomenclature is the term *policy*. Usage has become so slippery and contentious that some have accused the term of being meaningless. In an effort to help clarify it and related terms, the IETF Policy Framework Working Group created a document about policy that took over two years to complete. This document has been published as an informational RFC [RFC3198].

Merriam-Webster has several definitions for the term "policy." A common-sense one that works well for our purposes is:

> A definite course or method of action, selected from among alternatives and in light of given conditions, to guide and determine present and future decisions.

It turns out that this definition is essentially the same as the IETF's policy terminology document cited above. Crudely put, a network governed by policy is one in which decisions are dictated by a predetermined set of rules, or policies, instead of being made case by case. These policies could govern almost any network activity, from which networks' routing information will be accepted and shared, to how a particular type of traffic in the network will be treated. For example, priority might be given to traffic bound for the order processing systems. From these "policies," configurations appropriate to each network device that will enforce these policies must be set (see the discussion on a simple policy environment that follows). What "policy" means when it's modify-

---

*Some of the material in this section first appeared in articles previously published in a special edition on policy management in *Network Computing* magazine in the January 21, 2002 edition. They were co-authored with Bruce Boardman. The two relevant articles are: "No Standards, No Policy, No Management" and "Are you a Control Freak?".

ing "management" is…a whole lot harder to pin down. In the real world, policy management is an interesting concept to the extent that it can be leveraged to streamline and simplify network tasks. For example, instead of handcrafting the configurations for each of the network devices, a piece of software might take the policy, for example, one that gives higher priority to order processing traffic, and then generate the correct configuration appropriate to each device.

Implicit in the description above is a simple process. First, the policy is defined, then we select which systems are to carry out the policy, then we decide what configuration parameters each of the devices needs to cause the policy to be correctly carried out, then we send that configuration information to the devices. Later in our services management discussion, we will see that we must then monitor each of the devices to ensure that the policies have been carried out correctly.

The terms *policy-based configuration*, *configuration of policy*, and *policy-based configuration of policy*, which we discuss next, are helpful in understanding this process. However, keep in mind that in a practical way, we seldom think about these functions as separate, and the terms used here are for expository purposes.* When we use policy to configure devices and the services they support, we have a multifaceted problem. One part is the correct selection of the intended devices for configuration: policy-based configuration. Another part is figuring out the exact set of configuration parameters that must be sent to the device to cause it to perform the desired policy, such as treating some data on the network "better" or more securely than other data. The third element that ties the two previous points together is to match a specific device to a set of specific configuration parameters. The real challenge here is that devices and models differ, so we may have many specific sets of configuration parameters for different models and vendors, each producing the same policy behavior on the managed device.

Figure 4.1 illustrates the following terms:

- **Policy-based configuration**—The process of identifying the network elements that must be configured in support of a policy or policies. The policy can encompass any aspect of a device's operation, from DNS services to algorithmically assigned interfaces to certain types of services. An example of policy-based configuration could be

---

*These terms were discussed during the early stages of the SNMPCONF and were found to be helpful tools in understanding the general issues related to policy and policy based management.

Figure 4.1
The process of policy
configuration.

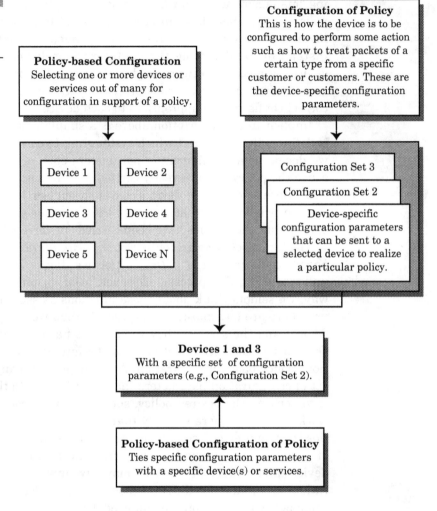

that all devices east of the Mississippi must use specific primary and secondary DNS servers.

■ **Configuration of policy**—The process of configuring devices to behave in conformance to a given policy, which is the specific configuration parameters that we send to a device (after they had been selected) and that cause the device to operate in conformance with the policy. A typical example of policy configuration in this sense is that we might send configuration parameters to a device that cause it to give all packets addressed to or from Jon Saperia preferential treatment. People often think about Differentiated services in connection with configuration of policy, but policy isn't just processing priority. It might be security-related, for example, as when a firewall disallows certain types of traffic on the basis of trust. In fact, policy can be just about anything related to the operation of devices and services on the network.

▪ **Policy-based configuration of policy**—As the unwieldy name suggests, this combines the two aspects of policy described above. In policy-based configuration of policy, the policy management system consults existing policies to determine what needs configuration and how it's to be configured.

Once again, remember that in normal operations these concepts are generally tightly coupled. We are being a bit pedantic in separating them here to illustrate the different aspects of a policy management system. The important point to remember is that policy in action is many different procedures, to some extent interdependent, and not readily summed up. Come to think of it, policy-based configuration *without* configuration of policy is exactly what happens without policy management software, and that's also an important point. Humans decide on a manual basis which machines will be used to realize a particular policy. Once the humans complete this step, they handcraft configurations for each and every type of system that implements the policy and then configure them separately. Here we can see the real benefit of a complete policy management system. If correctly implemented and deployed, it can automate a highly error-prone and labor-intensive process. At a minimum, it can help reduce errors when humans manually configure systems or write scripts for system configuration. My final point is that, when speaking with vendors, it is important to pin down what they mean when they say "policy."

Figure 4.2 is a diagram that will be familiar to many readers. It shows a network management station, or management application, in communication with the systems that it manages. The model dates from the earliest days of network management, and a Simple Network Management Protocol (SNMP) [RFC1067], is one of the early documents to reference this structure. It has long since been superseded by newer versions, but the basic principle has remained the same and it has not been said any better since then:

Implicit in the SNMP architectural model is a collection of network management stations and network elements. Network management stations execute management applications which monitor and control network elements. Network ele-

Figure 4.2
A simple policy
environment.

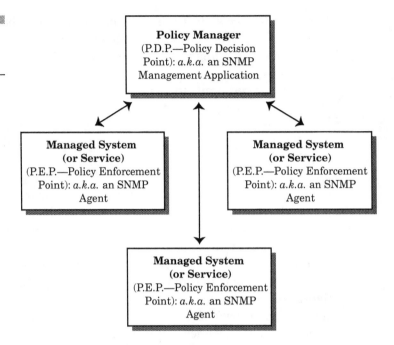

ments are devices such as hosts, gateways, terminal servers, and the like which have management agents responsible for performing the network management functions requested by the network management stations.

Even though we have new notions about management stations, agents, and functions generated from IETF working groups, other standards bodies, and vendors, the model hasn't changed much. A Policy Enforcement Point [PEP] and a Policy Decision Point [PDP]* are both defined by reference to configuration activities. A PDP sends configuration information to a PEP, which, once configured with the information, carries out (enforces) the policy. Each manager (or PDP in this case) controls many devices through its management agency (or PEP to use a new expression).[†]

---

*These terms are discussed in [RFC3198] Terminology for Policy-Based Management, and are generally helpful in understanding at least one way managers and managed devices can relate to each other in a policy management environment.

†For now we are sidestepping the case where multiple managers actively control a single device. That topic will be taken up in the requirements discussion.

So What's the Big Deal About Policy?

There isn't one. People have been using policy-based approaches to management for years. What *is* a big deal are some pertinent distinctions between how things were done in the past, and what the people who build policy-based management software today see as their goal, even if that goal is not always attained. In the past, if special configuration was needed, the task relied on a set of handcrafted scripts or "the guy who always does it." It was very much a black art. Now we have begun to formalize network configuration using policy-based applications and standards in place of wizardry. New protocols and ways of looking at information have been created specifically to support formal configuration. Products incorporating some of this technology have begun to emerge, and even traditional products have of late slapped a coat of paint on the old approaches and called them "new."

As we consider some popular technologies in the policy arena, stay focused on how they articulate basic concepts, instead of delving into specific protocols and data formats. We can take away from a discussion about policy these concepts, and apply them to the service management environment. These concepts include:

- Abstraction of data
- Abstraction of technical detail
- Adding device intelligence
- Modeling interactions and dependencies between network elements
- Propagating configuration information from a central policy repository
- Enabling dynamic reconfiguration

Abstraction of data improves the efficiency of data transfer. Many policy technologies espouse the principle that devices can be configured without knowing every configuration parameter for every configurable piece of the device. Proponents of this approach reason that many instances of a given configuration value on a device are identical, and so we can define configuration parameters as templates, not specific instances of values. Network device interfaces provide an example. As a rule, the parameters that must be supplied for a network device to operate correctly include the network mask, IP address, and default router, among others. Some, such as the IP address, are unique. In other words, configuration software must provide IP addresses to the network device for each and every interface. But that's not necessarily true for the network mask. There may be a single network mask value that is valid for many interfaces. In cases such as this, why send a value to a device twenty times over, when once is sufficient? We send a *template* value that can be instantiated on as many interfaces as is needed.

Abstraction of the details.   "Blackboxing" a technology means that a network can be run by fewer, less-skilled personnel. Most of us think that the more an operator knows about the intricacies of his system, the better he can manage the network. Generally speaking, this is true. The problem is one of scale: Where are we going to get enough people with the skills and knowledge necessary to run increasingly complex networks? Another principle of abstraction that typically benefits network operators is suppressing the details of how a technology works and how it is implemented from one vendor to the next. This makes it possible for a small number of experts to set things up and for larger numbers of less-skilled individuals to take care of the more routine configuration operations. For example, let's posit that three different bandwidth management technologies can control how much bandwidth the protocol for standard Web traffic is allowed to use a particular link. Unfortunately, each of the three has its own configuration and monitoring parameters, not to mention variations within each technology introduced by vendor implementations. Now suppose an operator wants to change a customer's bandwidth allocation over one or more links to put a new service into effect. As things stand, he must be conversant with the details of all configuration parameters for controlling bandwidth on each of those links just to get the job done—and that requirement constitutes a wasteful use of trained personnel to perform routine tasks.

The abstraction that we're talking about here is a type of *information hiding,* a well-known property of object-oriented software design. Object-oriented developers design programs so that one module can function without knowing how another module performs its function. The idea is to dumb down interfaces between software modules as much as possible. In dumb interfaces, each software module must learn to perform at most two tasks: the job it is designed to do, and the communication of messages to other software components. The same holds true in network operations: rather than focusing on the details of how each technology changes bandwidth, the network operator focuses simply on changing the amount of bandwidth for the Web traffic.

Information hiding has important benefits for policy. If the details of a network technology can be hidden from most of the people who must interact with it, we can deploy the experts to more-productive functions than the day-to-day configuration and management of these technologies in the network. Ultimately the key to this benefit of hiding the details of one or more technologies that accomplish the same or similar functions relies on also hiding the implementation details from one vendor to the next. Each vendor will implement the technology in a vendor-specific way. Standardization can be a big step toward information hiding, but it will never eliminate the natural tendency of vendors to implement technologies differently. In short, standardization in this area can help drive different vendors to be as common as possible in places where there is no operational benefit to differences. This is a win for everyone.

I'd like to stop and interject a proviso here. Don't take any of this to imply that vendor and technology experts can eventually be dispensed with. They will be needed to create the policies that govern the operation of network elements. That's a big part of the function they perform today, and policy designers developing management applications are making a concerted attempt to incorporate some of their smarts in ways that will be much more reusable than Perl scripts. Perl scripts, by the way, are usually created at the end-user site to do ad hoc per-element configuration. These scripts are not really suited to the type of management we have been discussing. We still have a long way to go before management software knows what a good (human) manager does, but we must go a long way farther before we can deploy more complex services effectively.

I've sometimes noted (to get into a philosophic diversion) that information hiding seems to be a natural tendency of technology. Consider the automobile, now in its second century of evolution: at first, automobile operators had to know how to adjust the fuel mixture and spark just to get the car started. Each manufacturer had different controls, so that operators couldn't transfer their lore from model to model, but had to relearn the ignition process each time. Assuming the operator was able to get her car running, she also had to know how to modify fuel mixtures for a variety of special conditions. Some early models even demanded manual adjustment of airflow over the radiator to warm up a cold car and cool down a hot one. You might say that management software today is not sufficiently removed from this primitive scenario.

Eventually systems evolved to chokes and carburetors and sophisticated cooling systems, and we began to hide some of these technologies. Now we have automobiles that not only do it all automatically, but cars with sufficient on-board intelligence to detect and report problems within these basic systems. Such problems are no longer relayed by the owner to the backyard mechanic; they read out directly to a computer display at the dealership. Each dealer's personnel have the knowledge and skills to cope with the problems of the makes and models they have been trained to service. You might say that this scenario describes the goal for modern management software. Clearly, the driver is now more dependent on his vendor than before and has to demand certain assurances with respect to equipment warranties and customer service, but he can do a great deal more with his car without having to know about the details of its inner workings.

Modern networks are in many ways more complex than modern automobiles, and they must operate in even more variable circumstances. More importantly, a car does not have any important operational relationships with other cars (except to avoid them), whereas network elements interrelate with many others, from their physical connection to their applications. These distinctions aside, the principles of abstraction apply in kind. It may be some time before data networks attain the degree of self-management cars now exhibit, but that's the job, at least to some degree, for service management applications. To realize the potential possible from policy management, they will have many of the attributes we describe in later sections.

Policies will require managed devices with more intelligence. Distributed intelligence enables a managed device to activate policy based on local configuration information sent by the management application. This distributed intelligence also includes the ability for the managed device to schedule the times when the policy is activated. These activation times are also sent to the managed device by the policy manager. Given that the system won't work without parameters for every knob, if we decide not to specify each and every one, but rely instead on a few defaults applicable to many instances of a parameter, just how is the system ultimately going to get configured? We give the device "rules" for applying these defaults. We can say, for example, that all Ethernet interfaces are to operate in full duplex mode. (For details on how it works, see the examples presented in the upcoming discussion of SNMPCONF technology.)

A well-considered and at least partially standardized model of how networks interact with their elements will facilitate policy. The concept of modeling networks has been around for quite some time but only recently has it been invoked in the context of policy-based management technology. By developing a clear understanding of how one function relates to another in a network or device, we can better understand how these devices should be configured and how to create better rules and abstractions to make configuration easier for network operators.

A "central" policy repository can be used to control the generation of configuration information throughout the network. Let's return to our discussion of policy-based configuration and configuration of policy: rather than having each person involved in configuration "remember" the details of how to configure each technology on each piece of vendor equipment, we can create a repository to store the policies (*policy-based configuration*) and the *configuration of policy* information—configuration parameters for a type of device and a technology. These stored policies can be used by the management system to create specific configuration information for each machine in the network that implements a given policy.

The possibility of dynamic reconfiguration. Many of the benefits of policy-based management are building blocks on which to create a dynamic provisioning system, but they aren't the whole story. The missing pieces needed to change configuration parameters on the fly will be discussed later in the chapter. In the meantime, administrators are justifiably uneasy when we start to talk about software systems that automatically reconfigure themselves in response to network conditions, because the work on these dynamic reconfiguration systems is just beginning. Correctly implemented, the policy concepts described in this section get us to the point where we can comfortably deploy management software that will amend operational parameters based on failures, utilization patterns, or other changes in the network environment. Elements of the network routing infrastructure already have these facilities and are trusted at least to some degree. Routing software can make decisions dynamically based on such factors as current load and connectivity. New products are emerging that look at the network and cause routers to change based on real-time observations of load, and also consider the dollar cost and other attributes of the lines between the routers. The network infrastructure of plain old telephone service (POTS) has long had features capable of resetting traffic patterns in response to failures.

What Drives the Need for Policy-based Management?

Operators are motivated either by cost containment or by a desire to create new high-margin services, and frequently by both. Service providers are investigating ways to get additional revenue streams from existing infrastructure, and enterprises are seeking ways to get more services out of last year's capital investment. In either environment, there are more knobs to turn, and they will probably have to be turned more often.

Part of the modern equation is that users demand more from networks. They want reliable service, reasonable assurance of security, predictable performance, and specialized services that never fail, such as videoconferencing. All of this adds to management complexity. The resulting networks are harder to configure; they have more interrelationships to troubleshoot and commensurately more ways to fail. Configurations on systems, particularly those at the edge, change more rapidly than they once did to accommodate new customers and their continual revisions in their selected services. Add to this the scarcity of skilled people to install, configure, and manage networks, and we have a real problem.

Historically, such a challenge was met by extremely adept individuals writing scripts to either touch the devices or to generate configuration files for them. This type of static provisioning may work well for stable devices at the core of a network, but it is more problematic in environments where change is more frequent and the complexity of the configuration operations is greater. Hence it's the edge that drives the need for policy-based management.

The bottom line is that policy-based management promises to facilitate the deployment of more complex services across a wider array of devices with fewer skilled workers. The result is increased revenue and decreased expenses—the new economics of the communications industry.

Which Technologies Are in Play?

So far, our discussion of policy has been rather abstract. We now look at some of the technologies proposed to advance the goals of policy-based management and, by association, the goals of services management.

Whereas every constituency for a technology claims to have made significant progress in terms of vendor interest, there is little evidence in the service management arena that one technology has or will gain dominance—at least in the near term. The technologies discussed here are often mentioned in connection with policy-based management. Keep in mind that this list is partial and summary in nature; it's intended to give readers a brief idea of what the technologies are and how they approach the larger problems we have been discussing. (You'll find pointers to additional information in the reference section.) Although

SNMPCONF is certainly relevant to policy-based networking, it won't be discussed until later in this section, because it offers the possibility of a solution for problems beyond policy and serves as the foundation for much of the discussion in later chapters.

In the following discussion, you will note that some of the technologies are not protocols for the movement of information between a managed device and the network or service management system (although some include protocols). Rather, some technologies focus on modeling information that is used by systems and the services that are built on top of them. In many respects, this is exactly the right focus: mechanisms (protocols) for the movement of management data are already plentiful—in fact, there are too many of them.

One of the central challenges to network and services management is to understand the relationships that exist between parts of a system and parts of a network. Some of the modeling proposed in these technologies describes some important relationships found in networks. Without this understanding, we cannot build an effective service management system. As an example, we need to know many different relationships to understand when an outage affects a service on a specific device and when it does not, or how services depend on each other.

Many of the technologies use object-oriented modeling approaches. For now, this is simply a methodology for expressing relationships. Later chapters will discuss some of the object-oriented principles that are important to modeling information for network and service management systems.

Common Information Model (CIM) and DEN

CIM is being developed by the Distributed Management Task Force [DMTF], which provides the following abstract of the CIM Specification on its Web site http://www.dmtf.org/standards/cim_spec_v22/.

> The DMTF Common Information Model [CIM] is an approach to the management of systems and networks that applies the basic structuring and conceptualization techniques of the object-oriented paradigm. The approach uses a uniform modeling formalism that together with the basic repertoire of object-oriented constructs supports the cooperative develop-

ment of an object-oriented schema across multiple organizations.

A management schema is provided to establish a common conceptual framework at the level of a fundamental typology—both with respect to classification and association, and with respect to a basic set of classes intended to establish a common framework for a description of the managed environment.

Much less formally stated, CIM uses object-oriented methods to standardize management constructs into a formal notation.

The management schema is divided into these conceptual layers:

* **Core model**—An information model that captures notions that are applicable to all areas of management.
* **Common model**—An information model that captures notions that are common to particular management areas, but independent of a particular technology or implementation. The common areas are systems, applications, databases, networks, and devices. The information model is specific enough to provide a basis for the development of management applications. This model provides a set of base classes for extension into the area of technology-specific schemas. The Core and Common models together are expressed as the CIM schema.
* **Extension schemas**—Represent technology-specific extensions of the Common model. These schemas are specific to environments, such as operating systems (for example, UNIX or Microsoft Windows).

Closely related to, and in some ways an extension of, CIM is the Directory Enabled Network [DEN] effort in the [DMTF]. Like CIM, DEN has made a real contribution to the way we think about network management data and its organization.

Given the emphasis this book places on the importance of data relationships and databases to service management software, it should be clear that a model like CIM offers real value. Although the models and the schemas might not be suitable for every environment, they do provide a thoughtful beginning for anyone building a management system or thinking about how to represent their network. This type of wide-ranging analysis of the objects associated with various aspects of the network management problem never took place in the IETF, and its omission has often proved problematic. The main problem has been that MIB Objects, the primary standardized management objects of the IETF, have had several difficulties.

The Management Information Base (MIB), which contains all the objects that are used by the IETF and extended by different vendors, is among other things, a model of data and their relationships presented in a different way from CIM or other approaches. Because data modeling is all about relationships, it is very helpful to understand as much of the entire "problem space" as possible before writing your management objects, whether they are expressed as CIM or MIB objects or anything else. The IETF's problem in this regard is that each group within the IETF working on a technology is responsible for creating its own management objects. Although group members may pay good attention to data relationships within that technology, they often do not pay as much attention to the details of other technologies. This is not a failure of SNMP or the IETF per se, just a side effect of the rapidly moving technology being developed in many different working groups. The priority of these groups is frequently not the manageability of the technology, but the base technology itself. Vendors also often extend the MIB with their private objects, which is a strength of SNMP. However, they often do this with little regard for existing objects and how well their system will fit together with other network devices from a management perspective. Any modeling technology will have this same problem of keeping a consistent model while at the same time allowing vendors to extend it. Another issue with MIB Objects is that the language in which they are written may need further extension. When the first version of the SNMP Structure of Management Information [RFC1065] was published in 1988, it met the needs of the community at that time. Since that time, it and networks have evolved. There is general agreement that to meet new management challenges—for example, dealing with the complex relationships in a services environment—SNMP must evolve further. It can do the job in its current state, but improvement would make the task easier.

These difficulties have made the development of management software significantly harder. Recently, the Policy Framework Working Group of the Operations and Management Area of the IETF has reacted by starting not only to identify useful terms (as we have previously discussed), but to work at creating a comprehensive model for management, one that has been lacking in the IETF until now. The Policy Core

Information Model—Version 1 Specification [RFC3060] is a first step. A great deal more work is required but this work, regardless of its outcome, is helpful.

Policy Core Information Model (PCIM)

The Policy Core Information Model—Version 1 Specification [RFC3060] takes the work of CIM and builds on it. The authors write:

> This document presents the object-oriented information model for representing policy information currently under joint development in the IETF Policy Framework WG and as extensions to the Common Information Model (CIM) activity in the Distributed Management Task Force (DMTF).
>
> In short, this beginning effort is part of an effort by the working group to not only define objects and their relationships in a more comprehensive way than has heretofore been accomplished. It is also an attempt to address issues of how one would model devices and services when one is concerned with policy based management.*

As with the original CIM work, PCIM and its derivative works add value as comprehensive views of how various network elements and services can work. Like CIM, PCIM and the various follow-on models suffer from a lack of concrete mappings to real network devices. The issue is that high-level models of services are very valuable: they can help us express behavior in a way that is easier for humans, rather than machines, to understand. The problem is that for these models to be useful in the real world, they must be tied to the models—however well or poorly laid out—that exist in the network equipment that is to deliver the service.

International Telecommunication Union (ITU)

The International Telecommunication Union [ITU] has been working on network management issues for many years and has contributed greatly to many commonly accepted concepts in the management domain. Although it has not been equally visible in the policy discussions, it has developed useful information models of network equipment from a telephony perspective. Two documents that are of interest here are a Recommendation and a subsequent amendment, [M.3100-1]. What is germane to our current discussion is that they first describe the importance of a model and its benefits, then go on to present at least a high-level view of important elements and their relationships in their environment. Modern IP networks are different in many respects from the traditional telecommunications network. For example, IP networks are based on datagrams as opposed to circuits. That said, there are some similarities; for example, the basic network elements and some of the relationships that should be considered when developing models of IP networks.

In the Overview of the Network Model that is described in Recommendation [M.3100] a key reason for developing a model is presented:

> A generic network information model is essential to the generation of uniform fault, configuration, performance, security, and accounting management standards. A common network model, identifying the generic resources that exist in a network and their associated attribute types, events, actions, and behaviors, provides a foundation for understanding the interrelationships between these resources and attributes, and

may, in turn, promote uniformity in dealing with the various aspects of managing these resources and attributes.

This basic principle informs many of the views in this book. At least as important as the understanding of the network, elements in the network, and services that are carried on the network, such a model enables the development of effective service management systems. The object relationships that are described later in this book not only foster an understanding of the problem of service management in an IP network, but are an essential building block for the development of effective service management systems.

Lightweight Directory Access Protocol (LDAP)

[LDAP] is frequently associated with policy work as a means of storing the policies used in a network in a standard way. If we are going to have network policies, they must be made available to management applications and often to the network systems themselves for use over and over again. Of course, network models that have been defined in CIM can also be usefully stored in these directories. Having a standard representation—in this case, a directory—for a network model, as well as policies in the network, is a very helpful principle. Some people have gravitated to LDAP as a method of storing policy information and network model information—it is a low-cost and relatively simple data storage technology—they have also encountered problems with it. The main problem, as we will see later, is not with LDAP itself or even with the attempt to create a common storage representation of policy information, which is a very helpful concept. The problem arises when we try to associate the policy information with current state and usage. That's when we come up against some of the inherent limitations of LDAP, such as its lack of performance when it comes to the rapid data insertion that is needed if we are going to collect and store policy usage information into the LDAP directory from polling many network objects frequently. It's possible to create a separate LDAP repository for policy information, and store the other usage and state data by means of relationship technologies. Although this will work, it makes for a more complex system since now we have to define, develop, and configure LDAP and a relational system. A preferable method is to use a relational system for all data, since relational technologies are able to not only store the policy information but can additionally be used to store usage and state data, along with most of the other information needed for a successful service management system.

The standards groups might have better served the community had they thought more holistically about the policy and service management problem and come up with a solution beyond simple policy storage. Paring the problem down to small chunks is often convenient and the shortest path to an initial release. This is what happens in product development, and when this approach is used in standards bodies, it has much the same affect—short-term gains at long-term expense.

Common Open Policy Service (COPS) and COPS-PR

The Common Open Policy Service [COPS] and COPS for Provisioning [COPS-PR] are two relatively new developments from the Resource Allocation Protocol (RAP) Working Group in the IETF. Many of the principles incorporated in these protocols are specifically for policy- and service-based management.* While any purpose-built protocol can have benefits over a generalized one, a new protocol carries with it a significant cost and obligation to deliver real value. A new protocol, whatever its benefits, carries costs associated with training, implementing, and debugging the new technology and understanding its use in real-world operation. So the question we must answer with regard to COPS and COPS-PR is: Does one or the other or both offer significant value over SNMP for their intended purposes? SNMP is a generalized protocol for management, whereas COPS and COPS-PR are more purpose built—in this case providing two approaches to configuration. As we will see in later chapters, multiple protocols make the management application development work much more difficult, and a single protocol such as SNMP might better serve our requirements. This is true even if SNMP requires some or a lot of improvement (depending on your perspective) to be an effective configuration management tool. COPS and COPS-PR were each developed for a different technology. COPS was initially deployed for Integrated Services (INTSERV), whereas COPS-PR was originally conceived as a COPS evolution to alleviate configuration prob-

---

*To review these technologies in more depth, please see the reference section for pointers to the relevant protocol specifications.

lems arising from Differentiated Services (DIFFSERV) configuration complexity.

In the case of Integrated Services, which was created to help address problems associated with real-time applications such as teleconferencing, it was found that a real-time signal was needed to reserve resources in a network to ensure that such applications work correctly. In turn, this signaling protocol, Resource Reservation Protocol (RSVP) needed to get information about resource policies so that it knew whether to grant the reservation request.*

The introduction to [RFC2748] reads in part:

> This document describes a simple query and response protocol that can be used to exchange policy information between a policy server (Policy Decision Point or PDP) and its clients (Policy Enforcement Points or PEPs). One example of a policy client is an RSVP[†] router that must exercise policy-based admission control over RSVP....

The overview document cited in the reference section was published in 1994, and many years later, the technology is still not widely deployed for its intended purpose. When Differentiated Services—a newer technology aimed at QoS problems in the Internet—was drafted, COPS-PR was created to send configuration information to devices that supported this new technology.

Differentiated Services had a least some goals that echoed those expressed in the mission statement of the Integrated Service documents. From [RFC2474], here is a brief summary:

> Differentiated Services enhancements to the Internet protocol are intended to enable scalable service discrimination in the Internet without the need for per-flow state and signaling at every hop.[‡] A variety of services may be built from a small, well-defined set of building blocks that are deployed in net-

---

*RFC2210 describes how RSVP is used with integrated services.

†RFC2749 describes how COPS is used with the Resource Reservation Protocol (RSVP) as a way of providing policy-based admission control over request for network resources.

‡A key difference between Integrated and Differentiated Services is that Integrated Services worry about per flow control for the reservation of network resources. The Differentiated Service technology does not worry about per flow control. Instead it looks at packets that have been marked in a particular way and treats packets so marked with a specific class of service.

work nodes. The services may be either end-to-end or intrado-main; they include both those that can satisfy quantitative performance requirements (e.g., peak bandwidth) and those based on relative performance (e.g., "class" differentiation).

By now you might be wondering what does all this alphabet soup of COPS, COPS-PR, RSVP, and Integrated and Differentiated Services have to do with our discussion. The essential point is that even if there is a growth in technologies for delivering different qualities of service in the Internet, it does not necessarily follow that it is a good thing to invent new management protocol for that technology, even if it can be optimized. COPS and COPS-PR are but examples of the overlap in capabilities of the management protocols. The generation of an optimized management protocol for a particular technology or function is understandable, because it can be tuned to the specific requirements of the technology it is to manage. That said, we should work hard to resist the temptation to develop these separate protocols, because Integrated and Differentiated Services are each managed with a different management protocol and environment. This raises the cost of deploying and developing management applications. While the two environments do have some similarities, they are different enough so that a management application would have to know the details of both COPS and COPS-PR in any system that had both Integrated and Differentiated Services.*

Our complaint here is not simply the proliferation of management approaches: it's that the proliferation is unnecessary. That's not to deprecate certain ideas applied in COPS and COPS-PR: for example, the COPS/COPS-PR method for managed devices to request or be sent provisioning information is much more concise than is current practice with SNMP and CLIs. In fact, that's an excellent idea. But instead of applying the new principles in COPS and COPS-PR to either SNMP or CLIs, which would have satisfactorily enhanced either one, two new infrastructures for management were developed, diverting time and resources away from solving real management problems. In general, when new technologies pose challenges to existing protocols that are functioning well in existing systems, new protocol development should not be considered the obvious solution. There are those who still feel that COPS and COPS-PR were inevitable because of defects in CLI approaches and SNMP; my view is that they were not needed.

---

*This seems unlikely given the current rate of adoption.

Extensible Markup Language (XML)

Given the intricacies of service management, it is not surprising that there is a regular stream of technologies proposed to solve some or all of the ills that confront management software. It is sometimes the case, as it is with the Extensible Markup Language [XML], that the technologies were developed outside the context of the management domain. A protocol does not have to be purpose-built to offer benefits to the management software environment: although XML was not intended to, and cannot, solve all the problems of management, it still has some important areas of applicability, one of which is the representation of configuration files for human consumption. Using XML, standard formats for configuration files could be developed across vendors right now, which would greatly aid operations experts in diagnosing configuration-related difficulties.

XML-based technology can play other roles in a services management system as adjuncts to other solutions. ASCII-based representations like XML are easier for humans to read than binary formatted data, and so many people would prefer that XML was the default method of transmitting configuration information between management systems and devices. This is an awkward approach because it seeks to optimize what should not be optimized: we want to optimize machine-to-machine interaction when sending configuration commands to a system, which is better done with machine-based APIs to standard machine-to-machine protocols and reasonably efficient, compact data on the wire. XML is a fine format to convert machine configuration data into ASCII readable forms such as when humans have to parse configuration information. But, just as modern word processing software does not try to store complex document data structures in ASCII form, neither should we reduce our management information to ACSII-only data. And just as a text version of word processing data should always be easy to generate, so too should XML formats for our configuration data.

Open Issues in Policy Management

Although policy-based management in general, and many of its allied technologies in particular, shed light on our study of service management software, it's currently unable, even in the aggregate, to deliver on the policy promise and still meet the full requirements of a service management system. While those technologies do carry key facilitating elements for service management, they are frequently conceived and implemented as islands of technology. In the sections that follow, we look at the shortcomings of policy technology as it is currently applied. In a sense, the major problem is that much of the current policy work is a useful increment—it just has not gone far enough to meet the requirements of a successful services management system.

Lack of Integration with Other Systems

For a service provider to offer high-value service and be profitable, it must be able not only to configure the services, but also to account and bill for them on a per-customer basis. Unfortunately, the vast majority of "policy-based" software currently available is configuration software. Once network devices are configured with policy managers, we do not have a clear idea of how to collect data from them so that accurate bills can be generated for individual customers. This obstacle is a result of two problems. First, most network systems do not have infrastructure that enables accounting based on configuration. Suppose we configure a particular interface that supports multiple customers to pass traffic of a certain type in preference to traffic of all other types? How will we then determine how much traffic of a certain type was passed on behalf of customer A and how much on behalf of customer B? The second problem is that policy management applications are not tied to any of the data collection systems commonly in use. We end up with a classic case of the right hand not knowing what the left is doing. In practical terms, we cannot accurately bill for sophisticated services even if we could create them in most modern IP-based networks.

Too Many Overlapping and Competing Efforts

An important distinction between how things have been done and how they'll be done in the future is that new products can leverage approaches that have not previously existed. These include standards for the expression of policy as well as protocols for conveying that information to systems. Let's look at several benefits of using standard protocols for policy-based management versus the traditional scripts that drive proprietary CLIs or in-house experts typing in the information on their own.

- A standard increases the likelihood that third-party developers will find it economically worthwhile to invest in tools that users can buy.

With standards, developers can provide for a wider range of products over a longer period of time, thereby getting a better return on their investment.

▪ As previously noted, policy-based protocols are designed for machine–machine communication as opposed to human–machine communication. Machine-to-machine protocols make it easier for developers to write the applications needed for effective policy management. They also make it simpler for administrators to manage many systems at one time—in short, they scale better.

The trouble with these new technologies is that there are so many of them. Some complement each other, while others overlap functions significantly and produce confusion in vendor and user communities. Vendors don't know precisely what to implement and customers are not always aware of what to ask for. Vendors often hedge their bets by implementing some of each technology, and, collectively, customers ask for everything. For application vendors, this nearly eliminates the advantages of standards just described. With many competing standards, the financial incentive for standards-based application development erodes. Users do not have an easier time of it, because they end up deploying multiple technologies to the same effect. Many find it better to remain with the tried-and-true approaches.

Take a closer look at one overlapping technology already identified. COPS-PR uses Policy Information Base Objects that are very much, but not exactly, like SNMP MIB Objects. Unfortunately relatively minor differences between objects create a wide disparity in practice. These differences extend to the method of conveyance of each type of object to the managed device. As we saw in an earlier discussion, incorporating good fixes into an extant framework such as SNMP solves problems, but creating a whole new infrastructure to institute a fix causes problems. The snail's pace of implementation and deployment suggests that the COPS-PR approach did not provide a critical mass of benefit to offset the cost of adoption.

Limited Range of Management Functions and Vendor Coverage

Most policy management systems have so far confined themselves to configuration functions. These policy management systems can help with policy definition and distribution in whatever form is appropriate to the managed devices. In the section "What Service Providers Require from a Service

Management System," we established that device configuration is just the beginning of the management challenge. Even within configuration, there are no generally accepted means for policy configuration across different vendors. Those vendors who claim to offer support do not do so across their entire product line. Some management applications attempt to work with equipment vendors' native interfaces by translating management data into proprietary, non-policy–based formats. This may help in the near term, but is costly to implement and hard to sustain, because a change from even one vendor can create the need for a software upgrade across the system.

The most significant limitation is that policy systems are not usually capable of collecting usage data associated with the policies they have set up. The data necessary to support billing or accounting for these services must be collected from other systems, or it is not collected at all.

In those cases where data is not collected by the systems that set up the configuration, we create a major roadblock to profitable service delivery, because a second system must be configured for data collection. For that to happen, configuration settings from the configuration system must be translated into information that will trigger collection of the right data.

Without this complex transformation it is difficult to generate accurate bills. Separate configuration and performance management systems also means that we have no fault data with which to verify service for simple service level agreements that guarantee uptime. True, we can try to rig it by attempting to collect these data with another system that can receive failure information and collect usage data, but even if we succeed it will be problematic to associate the data with the configured services set up by the policy system and link them to billable entities, such as corporate departments or ISP customers. In either of these cases, management systems simply pass the problem back to the network operator who, as a last recourse, must cobble some sort of system together or forgo high-margin services. The limitations described here affect even the simplest services currently offered by network operators.

Not Connected to Actual Systems

Earlier, we singled out the lack of a comprehensive modeling effort for MIB Objects on the part of the IETF—an oversight now being rectified.*

---

*With the work on the Policy Core Information Model and related information models, a considered analysis has begun in the IETF. See the Policy Framework Working Group Web page pointer in Appendix C for a current listing.

MIB Objects that represent things implemented in commercial products are often unrelated to the new models developed by one of IETF working groups. This is in part because we have over ten years of development in the MIB Objects in the IETF, and a corresponding history of development by vendors with their private MIB Objects. This, combined with the limited range of functions provided by these new approaches, makes associating configuration and usage data very nearly intractable, for the network operator. The other real reason for this "disconnect" is that these new models use new ways of defining objects that make it difficult to connect them to the many MIB Objects currently defined, thus exacerbating the problem of the connection of configuration data with usage information.

New Technologies Are Unstable

With the exception of SNMP, the technologies presented to this point are new. It can take years for a technology to move through the standardization process and make its appearance in a critical mass of network devices and management software. Such is the current state of policy management technology. With technology in a state of flux, it is not surprising that much network equipment today implements either proprietary technology or the interim versions of some standards. Developers of policy management applications thus are faced with the perennial problem of writing customized solutions that reduce their productivity and profitability.

Hype versus Reality

Hype is certainly not limited to policy or even networking products, but it's a particular problem in this area, where the need is great and the technology volatile. This is also an area where many alternatives compete for at least partial solutions. Vendors want to sell product so that they will be around to make product improvements. The creators of the various technologies are often aware of shortcomings and plan to fix them (eventually). With these high goals in mind, it is easy to see why the capabilities of products are sometimes overstated, at least for a particular product release.

Vendor Creativity

Another facet of the hype problem is that vendors like to differentiate themselves; this invariably means that they will create products that do not fit the standard models. Standard models are generally easily extensible, but as soon as vendors begin to diverge from them and from one another, the cost of developing an integrated management solution increases dramatically and the value of the standard declines.

Technologies for Policy Management

New technologies emerge at fairly regular intervals, so our short list is sure to evolve. The purpose of this section was to highlight some of the most relevant new technologies and provide a bit of context. By the time you read this, there are likely to be newer technologies that will claim to solve management software problems. You can expect them to have one or more characteristics to recommend them. The important question to ask as each one emerges is: Does the new technology outweigh the cost of adoption, or should we apply the productive concepts embodied in that technology to our existing management infrastructure? If the answer to that question comes out in favor of adopting new technology, the next critical question is: Does the new technology create overlap, even inadvertently, with approaches and protocols already in use? So far, we have not seen technologies that clear the bar on both accounts, although we've seen a great deal of clever thinking on policy problems. It now seems quite likely that no one of these technologies can solve all of those problems, and it is quite predictable that many more will come and go before the real benefits of policy management are realized.

I don't mean to trivialize the contributions of what has gone before. CIM, PCIM, and LDAP (among others) have all helped us organize our thinking about what network devices do, how to describe their relationships, and the services they can perform. COPS and COPS-PR are based on useful principles that demand our attention, such as finding a more concise way to send configuration information. XML also provides a structured way to look at configuration information, but XML is a long way from providing all the facilities needed for a generalized services management system.

Any incomplete solution must either duplicate facilities found in SNMP or integrate with them. A management system of any kind, whether it is for a set of elements, a network, or policy or service management system, must meet many requirements, which include facilities outside the scope of the new technologies we have listed. Examples of such facilities we've already discussed are data collection mechanisms and the means of transmitting error and state data—both essential to any reasonable platform for service management. So perhaps the third question to ask of any new technology contender is: Do we want it badly enough to undertake the painstaking job of integrating it into existing systems, or duplicating parts of existing systems to accommodate it? Either of these alternatives is unavoidably more costly than developing extensions for SNMP to incorporate the next great idea. *There are some who believe that the changes needed to make SNMP acceptable are so great that we would end up with an entirely new protocol. Although I do not subscribe to this view, even if we did change SNMP to create a new protocol, at least we would have one, and that is the important point.*

Configuration Management with SNMP (SNMPCONF)

A discussion of policy-related technologies would be incomplete without at least a brief discussion of SNMP and the new Configuration Management with SNMP [SNMPCONF] technology. Most aspects of SNMP-CONF are not new, but rather evolutionary, improvements to the Internet Standard Management Framework that uses SNMP. Over the years, this framework has shown itself to be a resilient platform for enhancement, and it is widely deployed as a fault management and data collection system. These properties make it an attractive platform to extend into the realm of policy and services configuration. In later chapters, we explore SNMP as a possible foundation on which to design, implement, and deploy an effective service management system.

SNMPCONF avails itself of all the existing mechanisms of SNMP and adds many of the concepts developed in policy-related technologies. It does this by describing new types of MIB Objects to act as templates that can be copied to individual configuration objects inside a machine. Consider a configuration parameter like the subnet example described earlier. In short, we've learned the inefficiency of sending exactly the same configuration information to each of the system's ten interfaces when those interfaces share the same value for an object, such as the subnet mask. We've learned that, in a management context, this kind of inefficiency is not just wasteful—it's actively inimical to holistic system management. Applying policy management means figuring out how to send one object once to ten different interfaces, instead of sending ten different objects one by one for each interface. When the object is sent, the local system can leverage policy by applying it to the ten interfaces that match the criteria for that particular value. That's the job of the Policy MIB Module [PM], developed to furnish objects for these rules.

The Venerable CLI

Did we really need a policy-based managemnt MIB module? Why can't we just use the CLIs supplied by most vendors of network equipment? Using CLI would save all this extra protocol work, and it's familiar to many in the user community. This is in line with all the principles of management software that we have discussed, and in fact, the CLI does have a role to play in service-based networks. In that role, however, it must coexist with other tools that are better optimized for management applications that are designed for less-skilled individuals and which perform more sophisticated functions than we can effectively create with a CLI.

Important Role of the CLI

CLIs are the human interface for a great many systems, and they have been optimized for efficient expert interaction with network devices. To use them at all, one has to know a lot about command structures, as well as each of the individual commands and the syntax for each. As we've repeatedly seen, the need for a human operator with knowledge of the machine is both a strength and a weakness. Recalling the auto analogy, a CLI can be equated to the diagnostic computer through which auto mechanics retrieve detailed information about how a car is running. In some cases, these interfaces also allow for changing values or parameters. For example, some cars keep track of service needs based on miles traveled and other operational conditions. The service technician can reset the system to start counting up to the next interval again when he has completed a service. Another good automobile example is the "check-engine" light familiar to many of us. We do not need to know the details of what is wrong, only that we need to bring the car in for service when the light goes on. The expert at the repair shop connects to the diagnostic interface in the car to read out information and determine the correct repair. In a sense, the CLI is optimized as a debugging tool using efficient, powerful commands that can change any aspect of a device's behavior. The CLI is the expert interface for making detailed corrections to a network device and for reading low-level information. It is the equivalent of "manual choke," the standard transmission, and diagnostic output in an automobile. The problem is that few drivers have all the skills necessary to interact with a car that has only these controls. Most of us want an automatic transmission, automatic choke, and simple read out (lights or simple gauges) for status or failure information. If we were to limit drivers' licenses only to those who had skills to run manual chokes, drive standard transmissions, and understand the computer error output, we would have far few drivers. (From some perspectives, a good thing—but definitely a problem if you have to get to work.)

A computer-related example of a CLI is the basic configuration of your desktop computer. Ten years ago, much PC configuration work was done either by editing configuration files or by issuing commands through the CLI. As more and more non–experts began using desktop computers, configuration interfaces evolved to accommodate their skills. Now it is possible to bring up a graphic user interface (GUI) that can accomplish many configuration commands at one time. In some cases, no application is needed at all—the computer simply does the work. Users do not have to know as much about the details as they used to. In addition, these interfaces often provide expert modes or editors for the same configuration files that have existed for years. CLIs are still present and available for experienced users. We need to preserve and improve our network CLIs in the same way, perhaps through standardization efforts or other means of making them more common across vendors. Additionally, CLIs properly enhanced, and taught a few policy and service management tricks, would improve efficiency and function.

Why a CLI Is Problematic for Policy and Service Management

At the risk of belaboring a point: the configuration of network devices is in many ways more complex than that of desktop devices, and complexity is the devil. First, network devices must be configured to interoperate with a number of other similar and, increasingly, dissimilar devices. Second, network boxes are becoming steadily more complex and will continue to do so in the future. Even if the complexity of each individual device can be controlled, the environment itself will become more complex. Network operators need machine assistance to configure a network capable of managing more sophisticated services.

The current generation of experts will always need and want the low-level debugging facilities that CLI provides. Future experts will also be needed, especially in more-elaborate networks. Are we likely to require these very high-level and expensive skills in the next generation of operators tapped to support the expansion of high-level value-added services in an increasingly complex network? Not all of them will come with the depth and breadth of experience characteristic of CLI operators today— nor need they. With or without expertise, they will be more productive with tools tailored to everyday service creation and management than with the tailored-for-the-expert interface of the CLI.

Text-based representations of configurations are sometimes essential to operators in the debugging process. These text-based representations can be used for a number of purposes, such as comparing one configuration file to the next to pinpoint where changes occurred or acting as a template from which another configuration file can be created for the same or different system. These types of representations are likely to be just as necessary tomorrow. Humans need ASCII representations to gaze into the computer. In the future, standardized formats for these files may be adopted, further sharpening our vision for debugging purposes. Unfortunately what is good for the human is not always good for the computer. Computers continue to get more and more sophisticated and powerful, but in the end, they always interact most efficiently with non-ASCII data.

Another dilemma for CLI-only and ASCII-only representations of configuration data bears mentioning: the more work we put into making the interface nice for humans in a CLI, the more work the network device has to do to translate the interface into something it can use.

I would argue that we always get the best results, no matter how nuanced our hardware gets, by optimizing the CLI as much as possible for humans and insisting on alternative canonic representations that are meant for machines. This is not much different from the dictum that we write computer programs in ASCII and compile them for machines. It's called the best of both worlds.

Finally, a CLI-style configuration is klutzy for policy and service management because the CLI user must log in to each system, perform the commands, and then log out and move on. We're all familiar with the procedure: log into the network component to be configured, make your changes, log out, and move on to the next. Yet most networks are not really configured this way. They use sophisticated programs that emulate the user. They log in, make changes, inspect values, and then log out. Scripts are also used to control the file transfer of configuration information. These programs and scripts place a thin veneer of automation on a simple approach. The basic model is unchanged though, and continues to be far less effective than an automated approach optimized for machine-to-machine communication.

For these reasons, a CLI can never be a great API for computer-to-computer communication, especially when we're trying for a one-to-many (manager-to-managed devices) relationship. Developing parsers and support code for ASCII-type interfaces is much more difficult when creating management application than it is when programming interfaces based on computer languages. An all-too-accessible proof of this point lies in the current state of configuration software for network devices: really good software is scarce and expensive to maintain and develop.

Policy and Service Management

Many of the issues of concern to policy are also the concern of service management software. For the purpose of our work, the primary distinction that exists between policy and service management software is the scope of work performed. Most policy software has, as a primary focus, configuration of one or more types of equipment, and we haven't seen much activity outside that focus. Little has been done, for example, to integrate policy configuration software with fault, performance, accounting, and security software (although a few vendors have begun to tie a few of the pieces together). Very little has been done to integrate policy management with the other business systems in an operator environment. All these levels of integration are mandatory for effective service creation, delivery, and management.

In the next chapter, we examine some of the abstractions that can be applied to integration efforts, all the way from the network equipment to business representations.

CHAPTER 5

Creating a Network Services Model

In this chapter, we discuss an interesting approach to thinking about networks that can help us better understand the requirements for services management software. In later chapters, we build on this understanding as we design and implement service management software. This chapter will also offer a practical background as you build and deploy services and management software in your network, or evaluate services and software systems for purchase.

I'll be using those concepts from the object oriented (OO) software engineering discipline that are particularly germane. These are not new concepts, as they relate to OO design, but here we apply them to the domain of service management in a consistent fashion across management applications, functions, and the devices and services they control.

Later, we diagram some of the principles as they relate to networks and the services that run on them, using the Universal Modeling Language (UML). Its graphic nature and increasing popularity help make these principles accessible to the widest possible audience. The UML used in this chapter is simple, but for those who want a fuller immersion, Meilir Page-Jones has written an excellent book [JONES].

Complexity of the Problem Space

Some people attempt to attack complex problems by breaking them down into small units. This is a useful technique, except that in some cases, the units must be dealt with as an integrated whole, making the problem complex again. According to Grady Booch:

> Developers who are content with writing small, stand-alone, single-user, window-based tools may find the problems associated with building massive applications staggering—so much so that they view it as folly even to try. However, the actuality of the business and scientific world is such that complex software systems must be built. Indeed, in some cases, it is folly not to try.

The networking community is faced with a similar challenge. For a long time we have had a small number of gifted individuals designing and configuring our networks. Unfortunately, the rapid growth in network size and complexity raises questions about the continued viability of this approach. For networks to run effectively, we must think of them as the large and complex systems that they have become. An organized examination of our networks is required, one that results in the creation of a generic network model, so that sophisticated software systems can be created to configure and control them.

In earlier chapters we reviewed some of the technologies and organizations that aim at modeling at least some portion of the network and its constitutive elements, including CIM and PCIM. Both tend to focus on one specific element type, such as a desktop device or router, or on one technology, such as routing or QoS, at a time. The purpose of the present discussion, which is very different, is to describe an overall model for a network and the services built on top of it. This is a prerequisite for being able to describe customer relationships to services and network elements. These important relationships enable us to build service management software capable of keeping the services running *and* documenting operations with a sufficient level of detail so that information can be output to accounting and billing systems accurately.

Object-Oriented Principles

In the previous chapter, I mentioned that the language used by SNMP to model its data might benefit from extension. The types of relationships that one can express with SNMP are limited in some respects and are not as easily represented as they are using the OO principles we discuss here and represented with UML diagrams* throughout the rest of the book. To build a service management system, we require a model from which to build the management software that can fully express the complex relationships and inter dependencies of services and customers.

The benefits of using OO design principles for management applications are well known, and a significant number of applications have been designed using at least some of these principles. Through time, application designers have found that these techniques have worked quite well. In this and later chapters, we use these principles to express new and more complex relationships than have previously been implemented in management software.

---

*To be more specific we are going to look at UML class diagrams. Many of the class diagrams will represent class hierarchies; however, the reader should not think that all class diagrams must be hierarchical.

We are not going to attempt any exhaustive review of all the object-oriented ideas that could be applied to the problem of understanding complex services in a network. Several good references are cited in this section for readers interested in more detail. Here we'll drill down only those basic principles that help in the initial organization of our thinking. These include abstraction, encapsulation, and hierarchy.

Abstraction

"Abstraction consists of focusing on the essential, inherent aspects of an entity and ignoring its accidental properties. In system development, this means focusing on what an object is and does, before deciding how it should be implemented" [RUMBAUGH]. For the most part, networkers have failed to apply this principle to our environment. Instead, we have tended to come up with a new control for every type of feature, model, or service we create, and we have done so even when there is considerable overlap in the essential properties of the controls. One reason we fail to question this sort of behavior is that it feels so natural. When we want to configure a service that gives low latency to certain types of traffic, we have no control that is the analog of a rudder on a plane. We must specify in great detail how the low latency is to be accomplished; this includes not only what we want the control to do but also how the control is to do the task. Grady Booch says:

> The discovery of common abstractions and mechanisms greatly facilitates our understanding of complex systems. For example, with just a few minutes of orientation, an experienced pilot can step into a multi-engine jet aircraft he or she has never flown before and safely fly the vehicle. Having recognized the properties common to all such aircraft such as the functioning of the rudder, ailerons, and throttle, the pilot primarily needs to learn what properties are unique to that particular aircraft. If the pilot already knows how to fly a given aircraft, it is far easier to know how to fly a similar one. [BOOCH]

Actually, I am pretty sure that in most commercial settings, the pilot would have to be certified on the new aircraft. But that's not the point—*abstractions*, such as the rudder controls, make it easier for a pilot to move from one airplane to the next. The more these controls differ, the more training is required.

To steer a plane, a pilot only has to pay attention to the steering control, not to all secondary characteristics of the system that influence the plane's direction. Our network operator does not have such an elegant control system. Instead of manipulating a rudder, to press our analogy, he has to identify all the hydraulic pressures that cause the plane to change direction, know how to set all of them, and set them all individually. Sound familiar? In a network, these "hydraulics" could take the form of queuing parameters that direct certain types of traffic to move more rapidly than other traffic. Because there are no high-level abstractions to work with, operators must deal with the low-level controls. The operator may further find that the low-level controls vary from one equipment vendor to the next. The same may be true of low-level avionic controls, of course, but the point is that the pilot doesn't need to know, and it's better that way.

Encapsulation

Closely related to abstraction is the property of encapsulation, which hides information about how something is implemented. "Abstraction and encapsulation are complementary concepts," says Booch. "Abstraction focuses upon the observable behavior of an object, whereas *encapsulation* focuses upon the implementation that gives rise to this behavior. Encapsulation is most often achieved through *information hiding*, which is the process of hiding all the secrets of an object that do not contribute to its essential characteristics...."

The principles of abstraction and encapsulation are particularly important for those of us working on network services management. If you write application programs, you will already be familiar with their value when applied to the general problems of software development for complex systems. The crux of this section, however, is that we start to apply these principles in the context of how we build out our networks and layer services on them. Of course, it's possible (although less likely) to have a great object-oriented design and implementation and still have a confusing human interface, so we want to use encapsulation and abstraction in building the interfaces to the network for our operations staff. An important benefit of doing so is that we can then create a language that is common across domains and easily approachable by more than just the experts in each. Right now, most of us can understand the terms "high" and "low latency." Fewer of us have mastered the details of how different parameters for various QoS technologies must be configured to create low latency. This principle of encapsulation can be most powerfully applied in the protocol-ridden management software domain. It lets us design a system so that, for much of the system software, the details of how each protocol does its work are sequestered from the software that needs information delivered or retrieved from managed devices. In practice this can be quite difficult, and that is one reason it is not often attempted, although, in the long run, this approach saves development time when more than one protocol must be supported.

Hierarchy

Many abstractions can exist in a network, and by placing them into a hierarchy we can better understand their relationships. When we want to discuss routing or QoS—abstractions—we may also want to talk about abstractions that are at a slightly lower level but still useful to the discussion at hand. Both Open Shortest Path First [OSPF] and Border Gateway Protocol [BGP] are types of routing services (protocols). The Integrated Services and Differentiated Services technologies are both QoS technologies. As it turns out, in networks and other technologies, there can be many levels in our abstraction hierarchy. Booch stresses the importance of selecting the right layer of a hierarchy when trying to identify problems, and so it is with networks and their services. A network operator might get an indication that a service is down—that's a service-level abstraction. At service level, an abstraction is unlikely to yield enough information to resolve the problem. For that, details further down the hierarchy will uncover what in particular is wrong.

Another important consideration with regard to hierarchy is that a complex system like a network will potentially have many different hierarchies in play. All of these hierarchies will fall into two main types: class and object. *Class hierarchies* help us understand common structures and behaviors, whereas *object hierarchies* help us understand how different objects interact with each other [BOOCH].

Let's climb back in the car for a moment. A simple class model of an automobile might have a drive train class that would include the engine, transmission, and wheels. An understanding of the drive train class hierarchy encompasses the various structures and behaviors of the drive system. We would know from an understanding of the wheel class, for example, that a wheel is a circular apparatus that moves forward or backward by rotation about an axis.* Understanding its object relation-ships—say, the relationship between the right front wheel and the right rear wheel—can have obvious benefits if you've got a flat tire and have to repair it. In this terminology, an object is a specific instance of a class; hence, the right front wheel is an instance of the wheel class. In the real world, it is never sufficient to know all the characteristics of an object's class; we also want to know its unique identity. In practice, the object and class hierarchies are interdependent, and we need to know about both to perform management work.

Figure 5.1 shows a simple abstraction hierarchy of our example automobile.

Figure 5.1
An automobile
hierarchy.

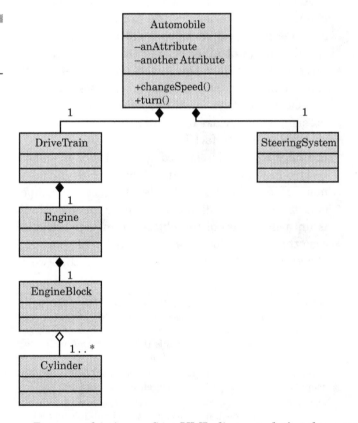

Because this is our first UML diagram, let's take a moment to investi-gate the notation. Each of the rectangles represents a *class*. "A class is the stencil from which objects are created (instantiated)." Therefore,

---

*A wheel class would probably have lots of other characteristics, but this is sufficient to make the point.

"each object has the same structure and behavior as the class from which it is instantiated" [JONES].

In Figure 5.1, the hierarchy could represent the automobile manufacturer's design. It is desirable that each car produced of a given model have the same characteristics as the next. Notice that each rectangle has three areas. The topmost, in bold letters, is the name of the class, and by convention the first letter is capitalized. The middle box represents attributes of the class, and the bottom box contains *methods* for the class. Methods are how instances of classes communicate with each other.* In this example, it might be a message to an automobile object telling it to change speed or switch direction.

Actually, speed change directives are a perfect example of showing a primary characteristic while hiding the implementation details. Our message to the car contains a desired speed, not the details of how that speed is to be attained. Using this hierarchy of abstraction, we do not have to know how the automobile changes direction; indeed the Automobile class *may not* implement steering functions. A message (in our case, the turn method) sent from the Automobile class to the SteeringSystem class in our diagram is an abstraction for all the details about steering. In fact it would encapsulate (hide) all the details of how the steering function is accomplished.

Two kinds of lines connect the classes in this diagram. A solid diamond means that the higher-level object is a composite of the lower-level objects. An outline diamond touching an object indicates that that object is an aggregation of the objects from which the line originates.† In our case the EngineBlock is an aggregation of potentially many cylinders, as shown by the '1..*' notation‡ indicating that we might create automo-

---

*Although we have not included them in this first diagram, methods can be used to perform tasks other than communicating with other instances of classes. They can be used to perform functions that are internal to the instance of a class. The most important point about methods for this discussion is that they represent the "behavior" of an instance of a class as viewed from the outside of that class.

†Note that the conventions for solid and outlined diamonds representing compositions and aggregations may vary in other reference books on UML.

‡The indication of how many cylinders might be in an engine block, in our case 1..* is fairly standard. In many diagrams, this indication of multiplicity could be at both ends of the line. If we were to have drawn a class for engineMounts, we might have a 1 at the engineMount end of the line indicating that an engineMount can only belong to one engine. At the engine end of the line (where the diamond would be) we might have a 1..*, indicating that we might potentially have many engine mounts on a single engine. I have skipped over the issue of whether this would be an aggregation or composition, since we would have to go into more detail about that diagram to make a correct determination. For the diagrams in this book, we will only put in these indicators of multiplicity where they are needed to help explain a concept in the text.

biles with one to many cylinders. Generally, an *aggregation* is made up of multiples of the same thing, for example, cylinders.* A *composition* is generally made up of different things, such as wheels, an engine, and an exhaust system [JONES].

We've explored these concepts in some depth to make a number of points, some of which will be expanded later:

- Graphic representation of object/class hierarchies can help our understanding of a complex problem.
- When designing a complex network and the services that will be layered on top of it, thinking about how different parts of the network and the services that run on it are related is important to help the design process of the network services, as well as software that is written to manage them.
- By using object-oriented principles, we can create better software that can be used to manage complex services more effectively by less-skilled individuals.
- By having a hierarchy that is connected from the most abstract business level, such as *Jon's traffic always gets gold service*, to the most concrete level, which will include specific configuration parameters sent to devices, we can avoid the discontinuities partial hierarchies suffer from. These discontinuities can introduce errors by humans as well as by the software, because either the people who perform mappings from one level to the next or the software must figure it out. Discontinuities can also greatly increase the cost of software development to manage the services. Also of importance when talking about discontinuities are those at the same level of abstraction. If, for example, we had a model for a PC or server that was fairly complete and none or only a partial one for the network devices that interconnected them, it would be difficult to manage services that used all of these components.
- The types of abstraction and encapsulation we have described can also reduce operator errors.
- A complete representation of the hierarchies is much more helpful than little bits and pieces. This incomplete representation is one of the criticisms of some of the technologies (e.g., CIM or PCIM) cited in Chapter 4.

---

*Note that the definitions I use in this book for composition and aggregation come from the references cited, and it is my goal to be consistent in the diagrams presented in this book. That said, there are others who have different meanings for aggregation and composition. What any reader of a UML diagram must be certain of is what the author intended.

Another important discontinuity in class and object hierarchies should be discussed in the management realm. This discontinuity is the result of the different access methods that people use to manage their network equipment. The two most common are SNMP and CLIs. As we have already demonstrated, SNMP does not use the modeling techniques we have just presented, but rather an organization of SNMP management objects called MIB Objects. In many systems, a parallel set of "objects" perform the same functions in a device when we type at the CLI. In some cases, some of the "objects" that exist in the SNMP hierarchy overlap with those in the CLI hierarchy. In other cases, they do not. In yet other circumstances, some objects exist in one hierarchy and not in the other. All of these discontinuities greatly add to the cost of the management software development: First, there is the cost of duplicate implementation, because often these systems do not utilize the principles of abstraction and encapsulation. Second, the management applications often have to deal with these different hierarchies, in which objects are named using different conventions. This is an important issue for both application developers and those who write management software for devices. (Both of these topics are discussed in greater detail starting in Chapter 6.)

Application to the Network

The following diagrams are an instructive way of looking at networks, services, and customer relationships. They help us establish the requirements for the design of the management applications and management software that execute on network elements.

Abstractions Relevant to Policy and Service Management

I've argued for the possibility that a system can be accurately represented by several different hierarchies. Indeed, complex systems may *require* many hierarchies to be represented fully. But simplification is still the driving force of abstraction, so each hierarchy will be founded on a subset of an object's most important characteristics.

Why is this so important in a networking context? As Booch says: "Abstraction focuses upon the essential characteristics of some object, *rel-*

ative to the perspective of the viewer" (italics mine), and we know that different classes of viewers have very different perspectives on the network.

In a service-enabled environment, many different constituencies (viewers) need access to information. Each viewer has his own, quite different perspective on what counts as "essential characteristics" of this information. Some may require a fair amount of detail, whereas others may only need to see the most highly abstracted information. (The pilot does not deal with the inner workings of the hydraulic system.)* On the other hand, in today's environment, people who plan to sell, monitor, and create additional instances of existing services in a network generally do have to know every detail of every technology.

Technology Hierarchies and Abstraction

The partial diagram shown in Figure 5.2 is a technology-centric view of the network. The hierarchical presentation, therefore, shows the abstraction and encapsulation of successively more-detailed information. Presentations of other sorts of network hierarchies—say, topologies for physical or logical characteristics—will look considerably different.

Note that in Figure 5.2 and others seen in this section, many of the method and attribute sections are either blank or fairly small, which would not be the case in a real network. The purpose of these minimal diagrams† is to uncover the basic ideas of hierarchies and their relationships in our network services model. In chapters on implementation, we will expand these and show some additional relationships between hierarchies. Figure 5.2 shows the hierarchical relationships between a service, domain, capabilities, and managed objects. An object that is a BusinessService is far more abstract than an object that contains information about how the BGP routing might be configured to deliver that business service.

---

*Failure conditions are an excellent example of the abstraction principles we have been discussing. In the event of a failure, additional information beyond the basic gauges might be required. Depending on the nature of the failure, an engineer might read voltages or go to service panels to take detailed readings. When the oil pressure light in our car comes on, we may check to see if there is enough oil by looking at the dipstick. If sufficient oil is present, then we take the car to an expert who will use lower-level diagnostics and tools to isolate and repair the problem. These tools are necessarily different from the tools that we see on a day-to-day basis.

†In this case, the multiplicity information has been left out. Were we to draw this diagram more completely, we might have indicated that all of the higher-layer services are compositions of one to many of the lower layer services. Many refer to the multiplicity attribute as *cardinality*.

Figure 5.2
A network services
hierarchy.

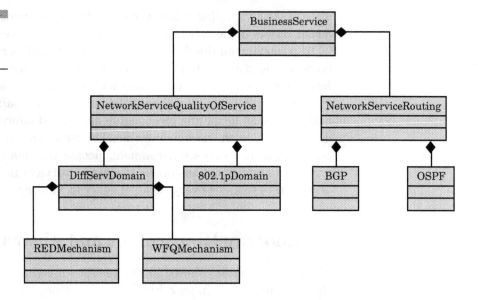

A business service represents the highest level of abstraction in our hierarchy. Information at this level is fairly simple and intended for consumption by people who may not possess a high degree of technical competency—customers, field sales reps, marketing personnel, and the like.

A business service is defined as a salable product. This could mean anything from high-reliability, secure Web hosting, to a set of corporate application servers that support interactive video training. As the diagram shows, these business services might well be comprised of many different network services. In this abbreviated example, we include two: QoS and routing. Note that there is a connection between the BusinessService and the NetworkServiceRouting and NetworkServiceQualityOfService classes. A business service in our example is built out of different NetworkServiceQualityOfService and NetworkServiceRouting objects. The connection is made explicit, because the BusinessService is made up of these other lower-level services. This is an essential point when it comes to keeping a network running and figuring out what is broken or how much service has been delivered: if the routing fails, so too does the business service. If the mechanisms that deliver the specific QoS features fail, the network may still be running, but the business service may not be providing the service at the agreed level of performance. To be effective, a complete service management system must be aware of these and many other relationships. This task is made all the more difficult because this is an area that has not been explored by standards organizations,* so each management system vendor is on its own with regard to how it supports this feature. It is left as an exercise for the reader to complete the remainder of the attribute and method information in the BusinessService class at the top of the diagram.

Suppose, for this example, that the network routing service is sold along with a certain latency guarantee for the traffic that is carried over the network. These two network services, routing and QoS, are combined to create a business service. The network services appear in the hierarchy as the NetworkServiceQualityOfService and NetworkServiceRouting classes.

The purpose of including this additional layer of abstraction is to support users who need to know that we have different technologies (domains) that might be used to realize a service. They neither need nor want to know the details of each technology contained at the mechanism layer (e.g., REDMechanism or WFQMechanism).[†]

NOC personnel want to know about whether the QoS levels are being met, but not how each knob at the mechanism layer[‡] is turned to create the desired latency characteristics. In fact, as our diagram shows, we might use several different domains of technology to deliver a single service. These other domains might be necessitated by differences in the physical media or, as in our routing example, whether the routing being performed is interior or external to the network.

Most network devices would not understand a simple command to set up Differentiated Services or BGP. These technology domains must be translated into the mechanisms that each uses to realize their tasks. Some mechanisms might be used by more than one service, and a complete model would reflect those relationships. In the mechanism layer of our hierarchy, all the details for each technology reside—the knobs that

---

*It is understandable that the standards bodies have not done much in terms of describing relationships between different types of network services. They have only just begun to describe the individual services themselves.

†The two mechanisms in this diagram that fall under the DiffServ domain do in fact represent technologies that are sometimes used by vendors that provide products that support differentialted services. For those interested in learning more, see the {RED} reference.

‡The mechanism layer that will be discussed in more detail in the next diagram is represented by the DiffServDomain and 802.1P domain classes, which are below the NetworkServiceQualityofService class and the BGP and OSPF classes that are used by the NetworkServiceRouting class.

we can turn and the counters we can monitor.* Here the "experts" in each technology are eagerly engaged. They know the details of how to configure a router with the Border Gateway Protocol (BGP), and they know how to monitor that router for correct behavior and performance problems. Nonexperts will content themselves with knowing if the routing protocol is operational, and thus will interact with our system at a higher level of abstraction. The systematic association of monitored values related to faults or performance with specific configuration details in our managed devices is an essential element of any model that can be used to help us build an effective service management system. Before we get deeper into that process, let's summarize the various levels of the hierarchy diagrammed in Figure 5.2:

- **Business Layer**—The layer that is most often used for provider-to-customer communications. When we speak of "gold" or "silver" service, often defined by some quantity of that service, we are at the business level of our hierarchy. Hidden below are all the details of how the service is created and managed.
- **Services Layer**—The services layer sits below the business layer but above the details of the domain and mechanism layers. At this level, we are not concerned with how a service is realized by certain technologies. We are concerned with specific services, however. QoS, routing, and name services are common examples.
- **Domain Layer**—The level at which we begin to specify particular technologies to realize a service. *Domain* is shorthand for the mechanisms used to execute a common task in the provisioning and ongoing management of a particular technology. It lets us know what technology is being used without knowing the details. In the routing example, BGP is one domain containing many mechanisms. In the case of directory services, at the domain level we would select from among the different technologies available to deliver directory services, either Lightweight Directory Access Protocol [LDAP] or [X.500].
- **Mechanism Layer**—A *mechanism* is a specific way to realize a service function. For example, in the service of routing, the BGP domain

---

*At this layer, the mechanism layer, the IETF and other standards bodies have been active over the years. More recent interest in higher layers has sparked activity in the IETF and other standards groups. Unless these layers are well connected to each other it will be very difficult to build the types of service management applications described in this book. The SNMPCONF approach, by using the same types of objects (MIB Objects) for all the layers and functions (i.e., FCAPS), makes it easier to build applications that can have the desired connectivity between these levels of abstraction.

has a series of attributes (mechanisms) that we use to understand (at least in part) the relationship of one router to another, a *peer*. The mechanisms include how often the two routers talk to each other to ensure that they are up—a *keepalive message*—or how often they exchange routing information.*

The four layers just identified from the hierarchy of Figure 5.2 are classes. Still missing are actual instances of the classes: essential components of the model, called *objects*. Discussions about service management are often about relationships between classes, instead of undertaking the much more conceptually difficult discussion of classes and the relationship of their instantiated values or objects. This concept must be followed by understanding how those instances are bound to real operation values in the network, and that is our next topic.

Classes and Instance Creation in the Network Services Hierarchy

Figure 5.3 illustrates the creation of objects from class definitions at various levels of the hierarchy. Instances of various objects are communicated to network equipment (not depicted in the diagram), where they can be applied or *localized* to parts of the device as determined by management software.†

The most important point to take away from this discussion is that a multilevel hierarchy supports software that can mask successive levels of detail from viewers who do not need it. At the same time it preserves the ability of technology experts to access data in a form that is most appropriate for their perspective.

---

*Some of these values are available in MIB objects and are found in [RFC1657] Definitions of Managed Objects for the Fourth Version of the Border Gateway Protocol (BGP-4) using SMIv2, S. Willis, J. Burruss, J. Chu, ed., July 1994. Unfortunately many of the mechanisms are only exposed via proprietary CLIs. This is an example of the problem we described when talking about access method discontinuities earlier in this chapter.

†Note that, particularly at the mechanism level, the binding of an instance of a class that has all the values filled in (localized) could be done either at the device or the network management system. If it is done at the network management system, we lose the efficiency gains associated with only sending the defaults that we have been discussing.

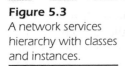

Figure 5.3
A network services
hierarchy with classes
and instances.

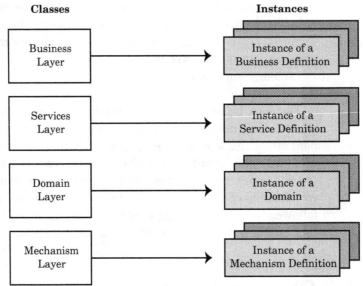

Figure 5.4 shows a business service built on a videoconferencing service and some routing services. (For simplicity's sake, I've only partially expanded the videoconferencing portion of the tree and have provided no detail for the routing portion.) Whether videoconferencing proves to be the next killer application for the Internet, it provides us with an example of how a hierarchy can be used to describe a network. Let's backtrack a little and see how it works.

From past discussions, we would expect functional business managers and executives to interact with and consume data from the business level. The business level of the hierarchy is quite simple and does not expose technical details of service delivery. It can and usually does include some of the following information:

* Customers who have subscribed to the business service, and details of when, where, and how the service is to be provided to each.
* How much of the service is already in use.
* How much more of the service the sales people can sell. This determination is based on additional information about the capacity of the network and servers and how much work these devices are currently performing. Here again, details of the capacity or usage are not exposed. All the sales people want to know is: Can they sell more service: yes or no?

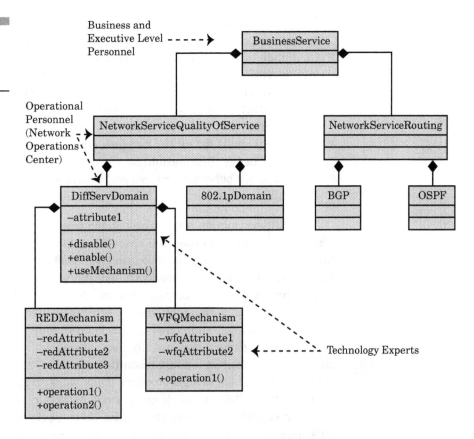

Figure 5.4
Mapping users to
the hierarchy—an
example.

※ History of customer accounts. Did they get what they paid for? Did they use more than they paid for? At the business level, such information may or may not include considerations like price, depending on the implementation. Many management systems are designed to output just enough information for accounting and billing systems to do their work.

Operational personnel will seek information about the ongoing health of the service and collect data about utilization from the next two levels of hierarchy; again, note that they will be less likely to tweak details at the domain and mechanism levels, although they may be aware of some of them. The operational personnel will generally be interested in:

※ State of the service and all hardware and software that realize the services. Three states are generally recognized as being useful: running, running in an impaired state, or not operational. Other state definitions can be created as needed.

* The provisioned state of the service and equipment. Are these devices and services supposed to be running, and if so, doing what?
* Customer relationships to the services and equipment—the data that lets operational personnel map failures to customers. This can help reduce the time from service failure to customer notification, a consideration that is becoming increasingly important in service-level agreements.
* Utilization levels for all the network elements.
* Customer usage information, on a per-service and per-device basis.
* Dependencies between services and equipment.
* Past failure information, to help resolve new failures.

The DiffServDomain and classes at the mechanism level* of Figure 5.4 contain made-up attributes and operations rather than actual names that might be used for these technologies. Our goal here is to show how this hierarchical structure can work, rather than getting lost in the details of each of the various technologies. In a real design for a DiffServ-based service, we might see various probability attributes used by whatever mechanism causes the system to treat one packet differently from another.

Technology experts are usually less interested in the current state of the service (except in a fault-resolution scenario) and much more interested in the service design problem. They want to know how changes in configuration affect network and service behavior. As we will see shortly, their expertise enables the higher-layer abstractions in the rest of this model. The information that these experts will extract from the mechanism layer includes:

* Exact values to configure each network device and service.
* Points of difference between models from a vendor. These differences range from varying performance capabilities when executing the same actions to varying configuration failure modes and configuration parameters.
* Points of difference between vendors. The same concerns that exist between different models from one vendor exist to an even greater degree between models from different vendors.
* Points of difference between access methods in use for the service (as we've seen in nonstandard CLIs).
* Interactions between all of these factors.

---

*REDMechanism and WFQMechanism are the classes at the mechanism level that support the (DiffServDomain) in this diagram. For those that wish to being an exploration of some of these mechanism, see [RED] in Appendix A.

The Problems of Localization

In this discussion, the *problem of localization* (some might call it binding) refers to the process whereby the value of an instance of an object in our hierarchy is copied to a specific instance of a configuration object in a managed device. This is a truncated definition. As we will see a bit later, localization can also work in other settings, as when a particular service definition is localized to a customer and for billing information.

This is a good place to introduce three terms: *default / template object*, *operational instance*, and *localization*. A default/template object (a.k.a. template) is an object used to fill in instances of configuration parameters in devices. These objects are intended for use by management system software, as opposed to the operational software in a device like a server or router. It doesn't matter if the management software is executing on a dedicated management platform or a network device.* Generally, IP interfaces are configured with a network mask in addition to an IP address. In many environments, devices with multiple interfaces use the same network mask on many or all of these interfaces, because that's how the network is designed. If a default/template object contains a value for a network mask, you may want to apply this value to the *netmask* object for some or all network interfaces in a network device. The management application sends this single default value to the managed device along with a rule about which interfaces the default is to be applied to. This reduces traffic, because the netmask value is sent only once to the network device instead of one time for each interface that needs the value.

An operational instance object is intended for use by network software or hardware to control or monitor network operation. Examples include specific configuration parameters for an interface such as an IP address or network mask.[†] In short, an operational instance object can

---

*In general, a management application has many default objects for a specific class. During the process of localization, the management software determines which of the many possible default values should be sent to the management software on a specific device, based on the device hardware, software, and other distinguishing characteristics. These default values, once loaded onto the managed device, are then used by the management software in a number of different ways including the creation of instances (copies) of the default that are bound to specific parts of the system and used by the operating software.

[†]In SNMPCONF, the word "element" identifies a physical or logical entity, such as a network interface. Elements can be thought of as resources that are used for service delivery. Each is represented in the SNMPCONF technology by a set of MIB Objects. Each instance of a MIB Object in a specific element is an operational instance object.

be created by making a copy of a default and saving it as an operational instance object for use by the operational software.*

Although nothing prevents the system from copying all kinds of objects in our hierarchy to an operational instance object, values for operational instance objects are almost always filled in with instances of default objects at the mechanism level of the hierarchy. This is because the higher one moves up the hierarchy, the more difficult it is for a network element such as a router or Web server to take these defaults and do something meaningful with them. In Figure 5.3, for example, the NetworkServiceRouting class at the Service Layer of Abstraction cites both BGP and OSPF as technologies to realize a routing service. That's certainly not sufficient information to give to a router and expect it to perform actual routing services. As anyone who has configured a router knows, a great many parameters relevant to the specific mechanisms used by BGP and OSPF must be fed to the router as well. And even this is not enough: the values for each mechanism for each technology domain must be associated with exactly the right operational instance objects (such as those found in network interfaces) in the managed device.

Localization is the method by which a higher-level default/template object is mapped to the next-lower level in the hierarchy, or ultimately to a specific operational instance object on a specific system.

Localization Details

We've discussed the information hierarchy and the types of information found at each level, and in this section, we need to "connect the dots." If we return to our videoconferencing example, what remains unexplained is how to translate the concept of a business service such as videoconferencing from one layer of our hierarchy to another, all the way down to specific values on specific systems in the network. This translation or *localization* has many dimensions. Several services could support the videoconferencing product, each of which may have different billing details or technology details that might be used to deliver the videoconferencing business service. Indeed, this localization process is highly problematic in the development of an effective service management system. The knowledge that is needed to translate the high-level concepts of videoconferencing to the

---

*In the SNMPCONF technology, the Operational Instance Objects are created by making copies of the defaults and applying these copies to the elements that match a selection rule called a PolicyCondition.

knobs that must be turned on a large number of systems usually only exists in the minds of the network experts. A service management system must be able to leverage these experts and cope with many dimensions of localization. We'll revisit localization concepts in later chapters. Here we just want to establish the importance of localization in a service management application. What follows are some general examples, along with a discussion of some of the factors that impact localization.

Creating Specific Defaults

Creating instances (objects) from the classes at the higher levels of abstraction can be thought of as creating defaults. Default creation, or localization, allows the details for each of the layers to be filled in. In Figure 5.3 we saw that each layer includes many instances, but we have yet to consider how those instances are generated. An effective service management system must have integrated facilities for this purpose, within certain limitations. No management system will ever be able to create defaults for all new services on the fly; instead, marketing and technical people cooperatively describe a business service and technical experts fill in the details for defaults as they move down the hierarchy, all the way to specific vendors, models, and even hardware and software configuration combinations. The process is called *service definition*; it's manual, and it won't brook shortcuts. Once created, however, defaults can be replicated by software and applied (localized) to any system that matches the selection criteria that can be evaluated by each machine.

Matching Businesses, Services, and Customers

If we have created our hierarchy correctly, deployed configurations properly, and set up the correct monitoring software, we still may not be able to offer a service that customers are willing to pay for. The final hurdle is *verification* of services. For verification purposes, it's not enough to know that a service instance is running or not running. We must associate a specific service with a specific customer and also connect this service instance with the network resources that provide the service. Without this association, we cannot verify that service level agreements have been met or that customers have overutilized a service.

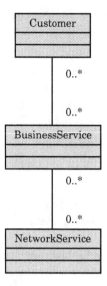

Figure 5.5
Mapping customers
to services.

Figure 5.5 shows the important relationship between customers, business services, and the network services that support them. If this diagram were fully fleshed out, it would have domains and methods as in previous diagrams. Here, we want to draw attention to some important details in the notation instead:

- The 0..* notation at the end of the line near the BusinessService class indicates that a customer may have zero to many business services, although it is not likely you'll ever see a customer with zero. Localization software allows users of the management system to match a specific instance of a business service with a customer. If we had drawn a complete object hierarchy, it would match an operational instance of the service with a customer.
- Depending on how the software is implemented, such a business service instance may be matched to many customers. The 0..* notation at the top of the line that extends from the BusinessService to the Customer class indicates that such a use is permissible.
- The same plurality is allowed between an instance of a BusinessService and a NetworkService. The implication for those who implement such systems is to build software that can distinguish the usage of a single instance of a network service by different customers, even though it might be easier to think of this if we have an instance of the service for each customer.

Selecting among many seemingly equivalent services at a higher level or customer perspective is part of the job of localization software. Where several network services are defined in very similar ways, and the differences between them reside in the details of technology domains and mechanisms, those differences could manifest themselves in terms of performance and reliability characteristics. The management software can only help the user in this regard if it can provide general information about these characteristics without requiring the user to know technical details. Monitoring software assumes that people who do know these details have set up the services and correctly localized the details at successively lower levels. To do so, they will have:

- matched the technologies to realize the service out of several potentially similar technologies;
- matched the right mechanisms to realize the technologies for each service; and
- matched the correct monitoring and usage objects to the service so that the service level received by the customer can be assured.

Matching Equipment and Capabilities

One of the most difficult tasks for localization software is to match the capabilities of the equipment to the desired services and the domains and mechanisms that realize them. Fortunately, this process can be automated. Software can be (and has been) written that can go out into the network and identify the capabilities of a specific network element. Work in the SNMPCONF area is adding additional facilities to this concept. The trick is to avoid attempting to configure a system that does not support a specific set of mechanisms with defaults that are meaningless to it. We can prime the management application with a list of preferences, so that if a particular piece of equipment does not support the top choice, but does support an acceptable alternative, the alternative defaults can be sent.

Matching equipment with capabilities is a multidimensioned problem, because it applies to equipment from many vendors, each with many models. In the worst case, it may require a separate set of defaults for every vendor and every model supported by each vendor for each technology and type of service in the network. Proponents of policy-based management have challenged this view, because one of the goals of policy management is to reduce the number of unique sets of defaults that must be created. Here, the elegance of the principle appears to be frustrated by the messiness of reality: there are just *so many* differences to accommodate from vendor to vendor and from model to model. As differences perpetuate, they tend to compound: higher-end systems have added features, and newer models evolve to support those features from the models already deployed. Clearly we will not soon see a network environment that runs on a single set of defaults, especially when that network has equipment from different vendors. I would suggest that the crux of the matter is not how many sets of defaults we use, but how many times each set of defaults can be used in a network and the process by which the defaults are created and applied. Where a set can be regularly recycled, the management gains are substantial. If your environment requires a custom set of defaults for each instance of each model of the equipment in use, your gain will be minimal.

Standard versus Proprietary Mechanisms

Some controls are standardized, whereas others are unique to one vendor, or even to a few models from a single vendor. In some egregious cases, a single device has been found to have duplicate controls for the same or related functions. The worse case of duplication occurs when a vendor creates a control that duplicates standard objects. This practice can cause confusion for operational personnel, and it can also raise the cost of management software that must deal with these overlapping objects. Localization should always be weighted in favor of choosing standard objects over proprietary ones. To make this choice, the software must understand each vendor and the range of private and standard objects available. Currently, SNMP has a much wider range of objects, both standard and private,* than any of the alternative technologies.

---

*The vast majority of standard objects are not for configuration; they are for monitoring and data collection. Some vendors provide configuration objects in their "private" MIB Objects. Recently, the IETF has provided more complete coverage of its protocols with SNMP configuration objects in addition to the traditional ones used for status monitoring and statistics.

Breadth of objects is a compelling reason for using SNMP as a basis for your service management system, as we will see in later chapters.

Localization of Defaults to Operational Instance Objects

For the most part, this function takes place inside a managed device. If we have a router with 100 interfaces, which of the hundred are to have a special set of defaults applied to them? The local system must have, or be provided with, the rules by which it can make the selection. Some technologies, such as COPS-PR and SNMPCONF, provide this capability today.

Figure 5.6 illustrates the process by which a device uses defaults, and the rules for their application, to identify where to apply them. The interfaces are those that have been selected in our example to have the defaults applied.

Figure 5.6
Selecting the operational instance objects.

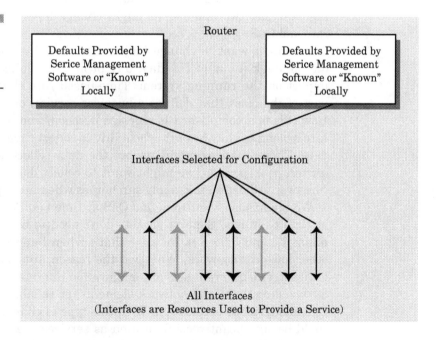

Localization at the Right Layer of the Hierarchy

In some cases, it may be necessary to determine whether abstractions above the mechanism level can be configured in a network device and if so, whether configuring them affects parameters in related mechanisms. If we have a system that allows routing services to be turned on and off, for instance, and we select off, the details for how BGP and OSPF are configured will be meaningless because they will have been placed in an "off" state.

Localization Based on Time and Utilization

Time is another dimension to the localization problem. Time is a factor in many telephone plans, in which the rate charged for certain types of calls varies with time of day or day of the week. In our videoconferencing example, two dimensions might change based on the time of day, rate and QoS.

We might want to change billing rates based on the time of day, or day of week, hour of the day, day of month. Doing so may have no impact on the running system. Time-based rate change could become part of the class that defines a business service, or part of a schedule. How this is accomplished is a design decision, and users will not much care as long as they have the flexibility to collect time-specific utilization data. These changes might impact the data collection system. During certain peak periods, we might want to collect data more often to capture over utilization and apply surcharges when appropriate.

We might also want to define QoS at time t_1 differently from quality at time t_2. Maybe Saperia packets don't need to be prioritized between midnight and 3:00 A.M. because that's when Saperia does backups or other low-priority work. Whatever the reason, time-based QoS requires (at least) two different sets of configuration information, either available for use directly by the network elements, or reliably sent to them from the management application. With this type of change in the mix, there could be significant reconfiguration as services are set up, modified, or removed. These changes occur at operational instance objects anywhere in our hierarchy.

Consider a service with utilization limits, such as a ceiling on the number of reduced-rate long-distance minutes in my telephone plan. This service might be defined so that I experience a service quality change when I've exceeded some limit during a specified time period. (There are bandwidth management products with this facility currently available for IP networks.) For such a system to work, it needs at least two sets of operational parameters. The first set for when I have not exceeded the limit, and a second set for when the limit has been exceeded.

Localization Discussion Summary

If the task of creating and running a system of this nature is daunting, consider the current conditions in most sizable networks. Each configuration must be either hand created, or created as the output of a specialized script. This is labor intensive, error prone, and still does not provide the types of associations that allow for accurate, time-dependent billing. Those faced with the responsibility for keeping the network functional have no "hooks" with which to rapidly determine the impact of a configuration change on the state of the network or the organization's profitability. Those responsible for generating bills are often unable to make the proper associations with configurations and billable entities. In a sense, the hierarchies and the types of localization so far described are a foundation on which we can expand from basic configuration to a complete management system.

Additional Dimensions to Our Hierarchy

Thus far, we have primarily focused on configuration information within the hierarchy. Although configuration is the foundation on which other types of management are based, by itself it does not enable a fully functional service management system. (This is also one of the deficiencies of policy management.) The hierarchies that we create for modeling a true service management system must include information for all the areas of management.

Configuration

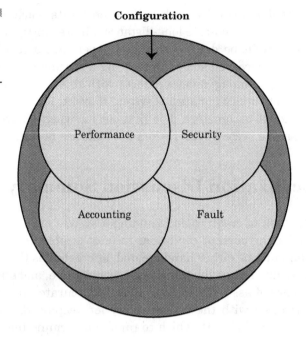

Figure 5.7 emphasizes the point that for effective service management we must create systems that facilitate the integration of all types of management data. From an implementation perspective, the question is how to integrate configuration into the model with all the other types of data. Let's look at the hierarchy we have been using, and extend what we have to incorporate these additional concepts.

In Figure 5.8, we have two familiar classes, NetworkService and Domain, plus two new ones, Capability and ManagementObject. With these additions we can demonstrate how a system could manage relationships between customers, devices, services, and management objects in devices using FCAPS information.

Starting at the topmost portion of the hierarchy represented in this diagram, we have added a few attributes to the NetworkService Class—customerList, deviceList, and domainList.

The customerList will be managed with appropriate methods so that we can assign and later discover which customers use a particular network service—information needed in the event of a service interruption for notification and billing adjustments. We might invoke methods such as associateCustomer. This method, when sent a proper message containing a customer identifier, can automatically add a name to the customer list for this particular network service.

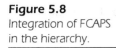

Figure 5.8
Integration of FCAPS
in the hierarchy.

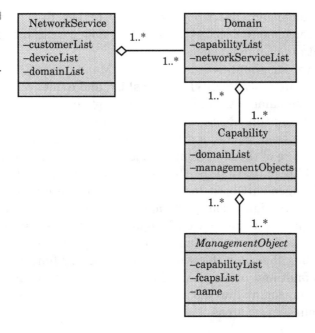

The deviceList exists to let us make analogous associations with the service, this time for network devices. Objects on the domainList are associated with network services. It makes little sense to attempt to configure a device to perform a function that it simply cannot perform under any circumstances. In the system we are describing, capabilities are used as a way of expressing what a system can and cannot do. Capabilities generally reside at the mechanism level of abstraction. In Figure 5.2, the service layer contains a service defined for routing, which has two possible technology domains—BGP and OSPF—capable of providing routing services. We could have expanded that diagram to include mechanisms that the BGP and OSPF domains use to accomplish their tasks. Capabilities may also be used to represent higher levels of abstraction.

Capabilities can help management software determine if a managed device supports a specific function required for a service. Because one capability might be used to support more than one domain, our Capability Class maintains and manages a list to keep track of the domains that use it. Realizing a capability may entail many pieces of management information; that information is held in individual management objects in our diagram. The notation 1..* indicates that, in our system, for a management object to be useful it must be connected to at least one Capability. In turn, Capabilities may require many

ManagementObject objects to fully realize their functions. Inside the ManagementObject we begin to see how we can integrate all the different areas of management together.

In our simplified drawing of a ManagementObject class are three attributes—capabilityList, fcapsList, and name.

The name attribute gives us a simple human handle to use when referencing this object. The capabilityList, in keeping with other objects we've seen, contains a list of Capability Object instances that use an instance of the ManagementObject for one or more purposes.

The fcapsList is the heart of the current discussion. Each ManagementObject instance has one or more entries in our fcapsList: fault, configuration, performance, accounting, or security. Normally, a single object can be used for multiple purposes, which saves us the extra data collection work (and storage and processing load) incurred by collecting the same object over again for every function that needs it, such as fault and performance. In this way, we can integrate all the areas of management into our hierarchy without creating one for each of the management areas.

Relationships between Management Objects

So far, we have not discussed relationships between Management Objects at the same layer of abstraction. Peer relationships can take place at almost any layer, as we will see in the design chapters that follow, and these mechanism-level relationships are particularly relevant in our current context.

If I experience a problem with certain pages on a Web server always turning up "not found," I would look at a predetermined place in the hierarchy for answers. Perhaps I would look at Web page management objects like those in [RFC2594], Definitions of Managed Objects for WWW Services. On the other hand, I might decide to check the status of mechanisms on which the Web server relies. The Web server probably relies on local operating system services visible through management objects in the System Application MIB Module [RFC2287]. Additional details might be found in management objects from the Application Management MIB Module [RFC2564[. Regardless of the source of the data, there is likely to be a hierarchy of dependency like that shown in Figure 5.9.

A real Web server has far more complex relationships than those depicted here, and it is correspondingly harder to model. Many services,

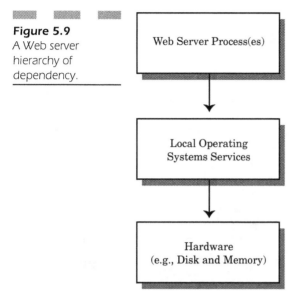

Figure 5.9
A Web server
hierarchy of
dependency.

such as DNS and routing systems, are subject to dependencies. To bill
correctly, management software must understand many relationships to
identify faults and other conditions and to generate bills correctly. In
the case of a user who has purchased a service and then finds it unavailable, it matters whether the failure comes from links in the user's home
system or from links in the provider network. In the case of faulty links
in the user's home, it is unlikely that the service provider will offer any
guarantees, although the provider does offer guarantees about the stability of its own infrastructure.

PART **3**

Creating
a Solution

O rganizing service management into a hierarchy with object oriented principles has a significant impact on how we organize our thinking about how to build service management systems. In Part 3, we discuss how management software can be created to meet the requirements already outlined and be consistent with the partial hierarchies already presented.

Chapter 6 begins our implementation discussion with an examination of SNMP as a basis for our service management system, and the properties that make it the best choice for that purpose. Chapter 7 looks at the work of an IETF Working Group that has proposed new technology to augment SNMP's ability to perform configuration management operations with the efficiency* so essential to a service management system.

While reading through Part 3 please keep in mind that there are two ways to think about the material presented. The first is as a set of specific recommendations for implementing a system, which may be helpful for some readers. The second and broader way is as a general approach to the problem of designing a service management system that can be adapted to different technologies, even if SNMP is not at the core. Those who evaluate, deploy, and use management software to help assess the software provided by others can use this broader view. So can those who want to assess an overall system so that deficiencies can be more readily identified and remedied.

My discussion of frameworks is presented mostly in terms of what the SNMP framework provides. If you choose to use a different framework, however, you should still consider the capabilities discussed in the context of SNMP. For example, the SNMP framework allows for access to management information based on who the user is, not just from where they are accessing the data. If you choose another framework, it should provide similar capability otherwise, it will prove hard to deploy, because most network operations personnel are allowed to perform management functions based on who they are or to which group of users they belong.

---

*Efficiency of information transfer is an important issue and one of the reasons that some of the other efforts we have discussed were proposed. As we will see, efficiency is only a small portion of the requirements for the platform on which to build a service management system.

CHAPTER 6

Designing a Service Management System

A service management system includes:

* all the management-related software that executes on managed systems in a network, and
* all the software on the managed elements that it interacts with—the *operational software.*

If the operational software has not been designed and implemented with manageability in mind, the software in the management system will not be as effective as it should be. As we have seen, the software that executes on managed devices is only part of the system: for a complete service management system, we must include the management applications as well.

[RFC2571] carried forward and expanded from earlier documents language that has been used for a number of years to describe what an SNMP-based management system is. This description is a good jumping-off point, especially since we will be using the SNMP framework for the design of our integrated service management system. To paraphrase these documents in simple terms:

> A management system will contain several and often times many network elements such as routers or servers that will contain management software that can send notifications. These devices will also be able to respond to requests for information, and accept commands as would be the case when configuring a device. To be complete, such management environments must also include at least one device that acts in the manager role. Minimally, it can receive notifications in the form of (asynchronous messages) from the managed devices, and can send commands as well as request information from the devices that it manages, and can process the responses.

Few of us are in a position to define all the software components for a system. Because different development groups and, in some cases, companies developed these components, we often have software components that do not work well together. Yet even with this limitation, the principles for building management software for both network elements and management applications still apply. The discussion in the following chapters will help you to make design choices based on these principles, and ultimately to develop software components that fit together better when deployed. As a minimum, you can take the information in these chapters and use what you wish as tools to evaluate existing systems or systems under consideration, even if they do not employ SNMP.

What Is a Framework?

The online version of the Merriam-Webster dictionary defines a framework as a basic conceptual structure. For our purposes, this definition is not quite sufficient. To design and build an effective service management system we also require building blocks that we can employ in different ways to create the variety of systems we need.

The Operations and Management Area of the IETF maintains a Web page with a number of references needed by those working with SNMP technology.* One of the most helpful is the boilerplate [BOILERPLATE] about the SNMP Framework. It is incorporated here, because we will be using the concepts contained in this material extensively. The boilerplate also offers a comprehensive set of references for those who wish to study the basics of SNMP in more detail:

1. The SNMP Management Framework
The SNMP Management Framework presently consists of five major components:

   ▪ An overall architecture, described in RFC 2571.
   ▪ Mechanisms for describing and naming objects and events for the purpose of management. The first version of this Structure of Management Information (SMI) is called SMIv1 and described in STD 16, RFC 1155 [RFC1155], STD 16, RFC 1212 [RFC1212], and RFC 1215 [RFC1215]. The second version, called SMIv2, is described in STD 58, RFC 2578 [RFC2578], STD 58, RFC 2579 [RFC2579] and STD 58, RFC 2580 [RFC2580].
   ▪ Message protocols for transferring management information. The first version of the SNMP message protocol is called SNMPv1 and described in STD 15, RFC 1157 [RFC1157]. A second version of the SNMP message protocol, which is not an Internet standards track protocol, is called SNMPv2c and

---

*See http://www.ops.ietf.org/ and the links from that page for more information. Also see the reference section at the end of this book.

described in RFC 1901 [RFC1901] and RFC 1906 [RFC1906]. The third version of the message protocol is called SNMPv3 and described in RFC 1906 [RFC1906], RFC 2572 [RFC2572], and RFC 2574 [RFC2574].

* Protocol operations for accessing management information. The first set of protocol operations and associated PDU formats is described in STD 15, RFC 1157 [RFC1157]. A second set of protocol operations and associated PDU formats is described in RFC 1905 [RFC1905].
* A set of fundamental applications described in RFC 2573 [RFC2573] and the view-based access control mechanism described in RFC 2575 [RFC2575].

A more detailed introduction to the current SNMP Management Framework can be found in RFC 2570 [REC2570].

Managed objects are accessed via a virtual information store, termed the Management Information Base or MIB. Objects in the MIB are defined using the mechanisms defined in the SMI.

This memo* specifies a MIB module that is compliant to the SMIv2. A MIB conforming to the SMIv1 can be produced through the appropriate translations. The resulting translated MIB must be semantically equivalent, except where objects or events are omitted because no translation is possible (use of Counter64). Some machine-readable information in SMIv2 will be converted into textual descriptions in SMIv1 during the translation process. However, this loss of machine readable information is not considered to change the semantics of the MIB.

To reduce this long description of the SNMP framework to its essentials, for a comprehensive management system to meet our requirements it must have the following facilities:

* A flexible and robust design for the transmission of management information.
* A comprehensive approach to security and access control.

---

*The reason that 'this memo' is in this boilerplate text is that the text is most often incorporated into new documents that extend the SNMP MIB and require this background information.

- Unique and extensible name space.
- Relatively simple protocol operations.
- A footprint inside the managed devices that does not significantly increase the memory or CPU resources of a well-designed system.
- Extensibility.
- A formalized "language" for the representation of the data.

Let's expand a bit on how these characteristics factor into good system design.

A Flexible and Robust Approach to the Transmission of Management Information

SNMP protocol specifications allow for its use over many different transports, although User Datagram Protocol (UDP) is required at a minimum. People have debated for years the wisdom of using such a connectionless mechanism. Operational results, however, show that UDP operates even when the network is under significant stress—exactly when you most need to control your network devices. Equally important for flexibility, SNMP provides a mechanism to enable alternative mappings to Transmission Control Protocol (TCP) or other transports. This facility has proved so desirable that there have been discussions about providing more support for it as part of the SNMP standard.

A Comprehensive Approach to Security and Access Control

SNMPv3 allows for three different levels of security and a way to control access to management information that is very similar to those systems that employ user IDs and passwords. In brief, when a system user identifies herself with an SNMP message, that message carries enough information so that the destination system can control what the user is authorized to see and do based on what group he is a member of. This concept is familiar to network operators, who are familiar with assigning engineers to distinct functional groups. One group may be primarily responsible for configuration and thus have read and write access. Another might be responsible for operational monitoring and thus have read access only.

A Unique, Easily Extensible Common Name Space

A common name space is a critical requirement for any management framework. In a "common name space," the naming conventions for management objects are the same for all objects, and each object is uniquely identifiable.* This requirement is nontrivial: for example, if a device has twenty interfaces, and each interface has five parameters, management software must be able to identify not only the specific parameter, but also the interface from which we want to get the parameter, (the instance) of interest. (See the section on Classes and Instance Creation in the Network Services Hierarchy in Chapter 5 for more information on this concept. In the SNMP world we can think of MIB Object definitions as loosely equivalent to the classes and instances of these in a managed device as operational object instances.) For example, the IETF has created a MIB object definition for the number of octets that an interface has received on a device. That could be the equivalent of a class. If we have a device with five network interfaces there would likely be five instances of that class, each with a value for the number of octets that interface has received since the management software that counts that information was initialized.

To be effective, a common name space must also provide a way for vendors to add extensions to the standards. This is valuable because vendor extensions can more easily reference standard objects if the vendor objects are created in such a way that they can use the same indexing as does the standard. For example, many standard MIB Objects exist for interfaces. As we saw earlier, we need to be able to identify the specific parameter (MIB Object) an instance of that object, to get the data we want. If a vendor uses the same index for the identification of the interfaces and adds its special parameter, software is able to show users the association between the standard objects that describe the status of an interface and the vendor extensions. The more standard objects in the system, the less costly it will be to develop management software.

In Figure 6.1, we see that objects created by the standards bodies and those created by vendors exist in the same name space. Therefore, at

---

*A simple analogy exists with the World Wide Web. We are all familiar with the format of a URL, a common name space for getting to web servers. Imagine how confusing it would be if some Web servers had to be addressed by the user with different combinations of hyphens and slashes and different prefixes other than "www" for example. The more difference there is, the harder for users and software to get work done.

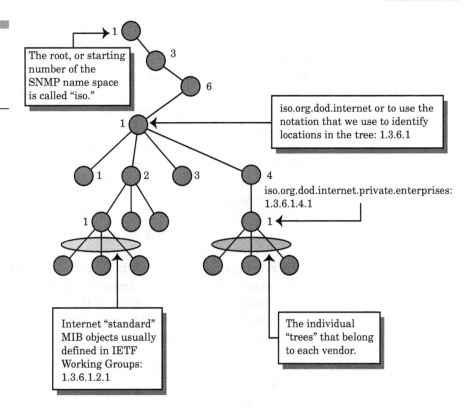

Figure 6.1

A common name space—standard and vendor-specific objects.

The root, or starting number of the SNMP name space is called "iso."

iso.org.dod.internet or to use the notation that we use to identify locations in the tree: 1.3.6.1

iso.org.dod.internet.private.enterprises: 1.3.6.1.4.1

Internet "standard" MIB objects usually defined in IETF Working Groups: 1.3.6.1.2.1

The individual "trees" that belong to each vendor.

some point, we have a common prefix. In the SNMP-based environment, which uses *object identifiers* (OIDs), that prefix is 1.3.6.1.

The shared prefix means that vendor objects can be understood in the context of standard objects. As an immediate benefit, we can create vendor objects that extend standard information. For example, over the years network interfaces have had many standard objects created for them, including counter and state objects. Via the standard index for the interfaces table [RFC2863], vendors can define objects, unique to their interface card, that describe aspects unique to their implementation. The ability to use the standard objects instead of creating new ones implies a substantial savings in development time. Management software that knows the standard objects only has to deal with the differences from vendor to vendor. By understanding the shared standard objects, the management application is implemented once for all products that implement these standard objects, whereas the management application must also have code that understands each of the vendors' unique objects if it is to effectively deal with these. More important, the "extra" objects, if implemented to share the same index in a MIB table with standard objects, can be understood to extend the standards objects.

Another benefit of the common name space is that it facilitates the integration of configuration with other types of management data, such as fault and performance. (Refer to Figure 5.7 to recapitulate that discussion.) In fact, the configuration, fault, and other data objects are all named within the same name space.

In Figure 6.2, the standard branch of the SNMP name space has a branch for Open Shortest Path First Routing Protocol [OSPF] Management information. This is noteworthy in our MIB Module discussion because it can be used to show the benefits of a common name space in several ways. One table contains objects for configuration as well as state and counter objects, thus it is possible to determine the state of a particular configuration, which is a critical component of successful service management. In addition, the OSPF Interfaces Table in the diagram uses, as one of its indices, the IP Address of the interface on which the OSPF area is running. Thus, this table is an extension of the *ipAddrTable*, and we can understand the information in the OSPF Interfaces table in the context of the *ipAddrTable*.

Figure 6.2

A common name space—configuration and other Object Integration

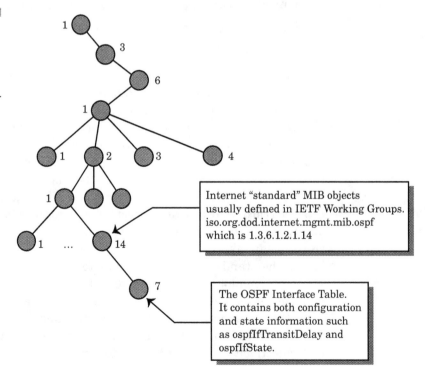

Internet "standard" MIB objects usually defined in IETF Working Groups. iso.org.dod.internet.mgmt.mib.ospf which is 1.3.6.1.2.1.14

The OSPF Interface Table. It contains both configuration and state information such as ospfIfTransitDelay and ospfIfState.

The OSPF Interfaces Table from RFC 1850

Since this is the first extensive inclusion of SNMP MIB Objects, a few explanatory words are in order. This and other extensive citations of MIB Objects are for the purpose of illustrating how extant MIB Objects can be used to help develop a service management system. We will not be concerned with the great number of details that one must worry about when creating new MIB Objects, that is a subject for other books. For those of you that are familiar with what MIB Objects look like, there is nothing new here. If you are not familiar with what these objects look like, browse through the included MIB Objects until you find the capitalized word "DESCRIPTION." Most of the objects that I have included have fairly detailed explanations in their DESCRIPTION clauses. Each is different, but if you read just these parts of the following citation and those that follow, you will get a sufficient understanding of what is going on without having to get too bogged down in all the rest of the details.

```
-- OSPF Interface Table

-- The OSPF Interface Table augments the ipAddrTable
-- with OSPF specific information.

  ospfIfTable OBJECT-TYPE
    SYNTAX   SEQUENCE OF OspfIfEntry
    MAX-ACCESS  not-accessible
    STATUS   current
    DESCRIPTION
      "The OSPF Interface Table describes the inter-
      faces from the viewpoint of OSPF."
    REFERENCE
      "OSPF Version 2, Appendix C.3 Router interface
      parameters"
    ::= { ospf 7 }


  ospfIfEntry OBJECT-TYPE
    SYNTAX   OspfIfEntry
    MAX-ACCESS  not-accessible
    STATUS   current
    DESCRIPTION
      "The OSPF Interface Entry describes one inter-
      face from the viewpoint of OSPF."
    INDEX { ospfIfIpAddress, ospfAddressLessIf }
    ::= { ospfIfTable 1 }
```

```
OspfIfEntry ::=
  SEQUENCE {
    ospfIfIpAddress
      IpAddress,
    ospfAddressLessIf
      Integer32,
    ospfIfAreaId
      AreaID,
    ospfIfType
      INTEGER,
    ospfIfAdminStat
      Status,
    ospfIfRtrPriority
      DesignatedRouterPriority,
    ospfIfTransitDelay
      UpToMaxAge,
    ospfIfRetransInterval
      UpToMaxAge,
    ospfIfHelloInterval
      HelloRange,
    ospfIfRtrDeadInterval
      PositiveInteger,
    ospfIfPollInterval
      PositiveInteger,
    ospfIfState
      INTEGER,
    ospfIfDesignatedRouter
      IpAddress,
    ospfIfBackupDesignatedRouter
      IpAddress,
    ospfIfEvents
      Counter32,
    ospfIfAuthType
      INTEGER,
    ospfIfAuthKey
      OCTET STRING,
    ospfIfStatus
      RowStatus,
    ospfIfMulticastForwarding
      INTEGER,
    ospfIfDemand
      TruthValue
        }

  ospfIfIpAddress OBJECT-TYPE
    SYNTAX  IpAddress
    MAX-ACCESS  read-only
    STATUS  current
    DESCRIPTION
      "The IP address of this OSPF interface."
    ::= { ospfIfEntry 1 }

  ospfAddressLessIf OBJECT-TYPE
    SYNTAX  Integer32
```

```
    MAX-ACCESS  read-only
    STATUS  current
    DESCRIPTION
      "For the purpose of easing the instancing of
      addressed  and addressless interfaces; This
      variable takes the value 0 on interfaces with
      IP Addresses, and the corresponding value of
      ifIndex for interfaces having no IP Address."
    ::= { ospfIfEntry 2 }


ospfIfAreaId OBJECT-TYPE
    SYNTAX  AreaID
    MAX-ACCESS  read-create
    STATUS  current
    DESCRIPTION
      "A 32-bit integer uniquely identifying the area
      to which the interface connects.  Area ID
      0.0.0.0 is used for the OSPF backbone."
    DEFVAL  { '00000000'H }  -- 0.0.0.0
    ::= { ospfIfEntry 3 }


ospfIfType OBJECT-TYPE
    SYNTAX  INTEGER  {
          broadcast (1),
          nbma (2),
          pointToPoint (3),
          pointToMultipoint (5)
          }
    MAX-ACCESS  read-create
    STATUS  current
    DESCRIPTION
      "The OSPF interface type.

      By way of a default, this field may be intuited
      from the corresponding value of ifType. Broad-
      cast LANs, such as Ethernet and IEEE 802.5,
      take the value 'broadcast', X.25 and similar
      technologies take the value 'nbma', and links
      that are definitively point to point take the
      value 'pointToPoint'."
    ::= { ospfIfEntry 4 }



ospfIfAdminStat OBJECT-TYPE
    SYNTAX  Status
    MAX-ACCESS  read-create
    STATUS  current
    DESCRIPTION
      "The OSPF interface's administrative status.
      The value formed on the interface, and the in-
      terface will be advertised as an internal route
      to some area.  The value 'disabled' denotes
      that the interface is external to OSPF."
    DEFVAL { enabled }
```

```
  ::= { ospfIfEntry 5 }

ospfIfRtrPriority OBJECT-TYPE
  SYNTAX  DesignatedRouterPriority
  MAX-ACCESS  read-create
  STATUS  current
  DESCRIPTION
    "The priority of this interface.  Used in
    multi-access networks, this field is used in
    the designated router election algorithm.  The
    value 0 signifies that the router is not eligi-
    ble to become the designated router on this
    particular network.  In the event of a tie in
    this value, routers will use their Router ID as
    a tie breaker."
  DEFVAL { 1 }
  ::= { ospfIfEntry 6 }


ospfIfTransitDelay OBJECT-TYPE
  SYNTAX  UpToMaxAge
  MAX-ACCESS  read-create
  STATUS  current
  DESCRIPTION
    "The estimated number of seconds it takes to
    transmit a link state update packet over this
    interface."
  DEFVAL { 1 }
  ::= { ospfIfEntry 7 }


ospfIfRetransInterval OBJECT-TYPE
  SYNTAX  UpToMaxAge
  MAX-ACCESS  read-create
  STATUS  current
  DESCRIPTION
    "The number of seconds between link-state ad-
    vertisement retransmissions, for adjacencies
    belonging to this interface.  This value is
    also used when retransmitting database descrip-
    tion and link-state request packets."
  DEFVAL { 5 }
  ::= { ospfIfEntry 8 }


ospfIfHelloInterval OBJECT-TYPE
  SYNTAX  HelloRange
  MAX-ACCESS  read-create
  STATUS  current
  DESCRIPTION
    "The length of time, in seconds, between the
    Hello packets that the router sends on the in-
    terface. This value must be the same for all
    routers attached to a common network."
```

```
                    DEFVAL { 10 }
                    ::= { ospfIfEntry 9 }


            ospfIfRtrDeadInterval OBJECT-TYPE
              SYNTAX  PositiveInteger
              MAX-ACCESS  read-create
              STATUS  current
              DESCRIPTION
                "The number of seconds that a router's Hello
                packets have not been seen before it's neigh-
                bors declare the router down. This should be
                some multiple of the Hello interval. This
                value must be the same for all routers attached
                to a common network."
              DEFVAL { 40 }
              ::= { ospfIfEntry 10 }


            ospfIfPollInterval OBJECT-TYPE
              SYNTAX  PositiveInteger
              MAX-ACCESS  read-create
              STATUS  current
              DESCRIPTION
                "The larger time interval, in seconds, between
                the Hello packets sent to an inactive non-
                broadcast multi- access neighbor."
              DEFVAL { 120 }
              ::= { ospfIfEntry 11 }


            ospfIfState OBJECT-TYPE
              SYNTAX  INTEGER  {
                    down (1),
                    loopback (2),
                    waiting (3),
                    pointToPoint (4),
                    designatedRouter (5),
                    backupDesignatedRouter (6),
                    otherDesignatedRouter (7)
                    }
              MAX-ACCESS  read-only
              STATUS  current
              DESCRIPTION
                "The OSPF Interface State."
              DEFVAL { down }
              ::= { ospfIfEntry 12 }


            ospfIfDesignatedRouter OBJECT-TYPE
              SYNTAX  IpAddress
              MAX-ACCESS  read-only
              STATUS  current
              DESCRIPTION
```

```
          "The IP Address of the Designated Router."
       DEFVAL  { '00000000'H }   -- 0.0.0.0
       ::= { ospfIfEntry 13 }


  ospfIfBackupDesignatedRouter OBJECT-TYPE
     SYNTAX   IpAddress
     MAX-ACCESS  read-only
     STATUS  current
     DESCRIPTION
       "The IP Address of the Backup  Designated
       Router."
     DEFVAL  { '00000000'H }   -- 0.0.0.0
     ::= { ospfIfEntry 14 }

  ospfIfEvents OBJECT-TYPE
     SYNTAX   Counter32
     MAX-ACCESS  read-only
     STATUS  current
     DESCRIPTION
       "The number of times this OSPF interface has
       changed its state, or an error has occurred."
     ::= { ospfIfEntry 15 }


  ospfIfAuthKey OBJECT-TYPE
     SYNTAX   OCTET STRING (SIZE (0..256))
     MAX-ACCESS  read-create
     STATUS  current
     DESCRIPTION
       "The Authentication Key. If the Area's Author-
       ization Type is simplePassword, and the key
       length is shorter than 8 octets, the agent will
       left adjust and zero fill to 8 octets.

       Note that unauthenticated interfaces need no
       authentication key, and simple password authen-
       tication cannot use a key of more than 8 oc-
       tets. Larger keys are useful only with authen-
       tication mechanisms not specified in this docu-
       ment.

       When read, ospfIfAuthKey always returns an Oc-
       tet String of length zero."
      REFERENCE
       "OSPF Version 2, Section 9 The Interface Data
       Structure"
     DEFVAL  { '0000000000000000'H }   -- 0.0.0.0.0.0.0.0
     ::= { ospfIfEntry 16 }

  ospfIfStatus OBJECT-TYPE
     SYNTAX   RowStatus
     MAX-ACCESS  read-create
     STATUS  current
```

```
DESCRIPTION
   "This variable displays the status of the en-
   try. Setting it to 'invalid' has the effect of
   rendering it inoperative. The internal effect
   (row removal) is implementation dependent."
::= { ospfIfEntry 17 }


ospfIfMulticastForwarding OBJECT-TYPE
   SYNTAX  INTEGER  {
            blocked (1),    -- no multicast forwarding
            multicast (2),  -- using multicast address
            unicast (3)     -- to each OSPF neighbor
        }
   MAX-ACCESS  read-create
   STATUS  current
   DESCRIPTION
      "The way multicasts should forwarded on this
      interface; not forwarded, forwarded as data
      link multicasts, or forwarded as data link uni-
      casts.  Data link multicasting is not meaning-
      ful on point to point and NBMA interfaces, and
      setting ospfMulticastForwarding to 0 effective-
      ly disables all multicast forwarding."
   DEFVAL { blocked }
   ::= { ospfIfEntry 18 }


ospfIfDemand OBJECT-TYPE
   SYNTAX  TruthValue
   MAX-ACCESS  read-create
   STATUS  current
   DESCRIPTION
      "Indicates whether Demand OSPF procedures (hel-
      lo supression to FULL neighbors and setting the
      DoNotAge flag on proogated LSAs) should be per-
      formed on this interface."
   DEFVAL { false }
   ::= { ospfIfEntry 19 }


ospfIfAuthType OBJECT-TYPE
   SYNTAX  INTEGER (0..255)
          -- none (0),
          -- simplePassword (1)
          -- md5 (2)
          -- reserved for specification by IANA (> 2)
   MAX-ACCESS  read-create
   STATUS  current
   DESCRIPTION
      "The authentication type specified for an in-
      terface.  Additional authentication types may
      be assigned locally."
   REFERENCE
```

```
      "OSPF Version 2, Appendix E Authentication"
    DEFVAL { 0 } -- no authentication, by default
::= { ospfIfEntry 20 }
```

Many of the objects in this table relate to the basic configuration of OSPF. In particular, if a router is having a problem with OSPF, you might want to investigate the values in this table and other OSPF objects. By virtue of the fact that the table uses a common name space with, and indeed shares the same index value as, the IP address table, you'll see other potentially relevant information from the IP address table that could impact the OSPF behavior. The common index values in the above OSPF Interfaces Table and the IP address table thus allows us to "connect" the OSPF Interfaces Table with other tables indexed by interface number (we'll talk about how this works later). The *ipAddrTable* from [RFC2011] is formed as follows:

```
-- the IP address table

ipAddrTable OBJECT-TYPE
  SYNTAX    SEQUENCE OF IpAddrEntry
  MAX-ACCESS not-accessible
  STATUS    current
  DESCRIPTION
      "The table of addressing information relevant to this
      entity's IP addresses."
  ::= { ip 20 }

ipAddrEntry OBJECT-TYPE
  SYNTAX    IpAddrEntry
  MAX-ACCESS not-accessible
  STATUS    current
  DESCRIPTION
      "The addressing information for one of this entity's IP
      addresses."
  INDEX    { ipAdEntAddr }
  ::= { ipAddrTable 1 }

IpAddrEntry ::= SEQUENCE {
    ipAdEntAddr       IpAddress,
    ipAdEntIfIndex    INTEGER,
    ipAdEntNetMask    IpAddress,
    ipAdEntBcastAddr   INTEGER,
    ipAdEntReasmMaxSize INTEGER
  }

ipAdEntAddr OBJECT-TYPE
  SYNTAX    IpAddress
  MAX-ACCESS read-only
  STATUS    current
  DESCRIPTION
```

```
              "The IP address to which this entry's addressing
              information pertains."
          ::= { ipAddrEntry 1 }

    ipAdEntIfIndex OBJECT-TYPE
      SYNTAX    INTEGER (1..2147483647)
      MAX-ACCESS read-only
      STATUS    current
      DESCRIPTION
          "The index value which uniquely identifies the interface to
          which this entry is applicable. The interface identified by
          a particular value of this index is the same interface as
          identified by the same value of RFC 1573's ifIndex."
          ::= { ipAddrEntry 2 }

    ipAdEntNetMask OBJECT-TYPE
      SYNTAX    IpAddress
      MAX-ACCESS read-only
      STATUS    current
      DESCRIPTION
          "The subnet mask associated with the IP address of this
          entry. The value of the mask is an IP address with all the
          network bits set to 1 and all the hosts bits set to 0."
          ::= { ipAddrEntry 3 }

    ipAdEntBcastAddr OBJECT-TYPE
      SYNTAX    INTEGER (0..1)
      MAX-ACCESS read-only
      STATUS    current
      DESCRIPTION
          "The value of the least-significant bit in the IP broadcast
          address used for sending datagrams on the (logical)
          interface associated with the IP address of this entry. For
          example, when the Internet standard all-ones broadcast
          address is used, the value will be 1. This value applies to
          both the subnet and network broadcasts addresses used by
          the entity on this (logical) interface."
          ::= { ipAddrEntry 4 }

    ipAdEntReasmMaxSize OBJECT-TYPE
      SYNTAX       INTEGER (0..65535)
      MAX-ACCESS   read-only
      STATUS       current
      DESCRIPTION
          "The size of the largest IP datagram which this entity can
          re-assemble from incoming IP fragmented datagrams received
          on this interface."
    ::= { ipAddrEntry 5 }
```

To complete this picture, note that the two tables just presented are related to the interfaces table found in the Interfaces Group MIB [RFC2863].

Before we look at the Interfaces table, let's look at the index objects for these tables to make clear how they are connected from the OSPF Interface Table:*

```
ospfIfIpAddress OBJECT-TYPE
    SYNTAX      IpAddress
    MAX-ACCESS  read-only
    STATUS      current
    DESCRIPTION
      "The IP address of this OSPF interface."
    ::= { ospfIfEntry 1 }
```

The handle here (the way we specify exactly which instance) is the IP address of the network interface that is being described by an entry in the OSPF Interface table. There may be many such entries (rows)[†] in this table, so let's see how this is connected to the IP address table. The index, or handle that one would use to get information about an IP address on a device from this table is via the *ipAdEntAddr* index object:

```
ipAdEntAddr OBJECT-TYPE
   SYNTAX      IpAddress
   MAX-ACCESS  read-only
   STATUS      current
   DESCRIPTION
      "The IP address to which this entry's addressing
      information pertains."
   ::= { ipAddrEntry 1 }
```

This IP address value is the same as that described in the OSPF Interface Table, thus, this is how these two tables are connected. If we know an IP address on a device, we can get more information about it from the IP address table as well as any OSPF information (if OSPF is running on the interface that has been assigned the IP address in question).

Notice that in the IP address table there is an object:

```
ipAdEntIfIndex OBJECT-TYPE
   SYNTAX      INTEGER (1..2147483647)
   MAX-ACCESS  read-only
   STATUS      current
   DESCRIPTION
```

---

*Careful readers will see that the OSPF Interface table has another object used to help with interfaces that do not have addresses. For the current discussion, we are leaving this out for simplicity.

†Each row contains instances of objects like counters or configuration parameters for a specific interface.

```
                   "The index value which uniquely identifies the interface to
                   which this entry is applicable. The interface identified by
                   a particular value of this index is the same interface as
                   identified by the same value of RFC 1573's ifIndex."
             ::= { ipAddrEntry 2 }
```

This object holds the index value of the interface that has been assigned to the particular IP address in question. This is how this table is connected to the next table, the Interfaces Table, which is indexed by the interface: the value of *ipAdEntIfIndex* must be the same, according to these specifications, as the value found in from the Interfaces table:

```
ifIndex OBJECT-TYPE
    SYNTAX        InterfaceIndex
    MAX-ACCESS    read-only
    STATUS        current
    DESCRIPTION
        "A unique value, greater than zero, for each interface. It
        is recommended that values are assigned contiguously
        starting from 1. The value for each interface sub-layer
        must remain constant at least from one re-initialization of
        the entity's network management system to the next re-
        initialization."
    ::= { ifEntry 1 }
```

As you will see from the Interfaces table that follows, there is a great deal of information that would be of value to an operator trying to find out why other routers in the network weren't able to share information with OSFP in a given device. By looking at the three tables, the operator could see OSPF information and the specific interface over which it should be communicating with those routers that have lost contact. If the interface is down, this tells the operators one thing: no communication with other network devices can take place over that interface—it is down; not operating. If, on the other hand, it is up and running, then they must look further for the problem. They might look at the OSPF information in the table previously presented to help isolate the problem.

I've gone into such detail on this point to emphasize the importance of the relationship of operational state data, such as that found in the Interfaces table, with configuration data, such as that found in the previously presented OSPF Interface Table.

```
-- the Interfaces table

-- The Interfaces table contains information on the entity's
-- interfaces. Each sub-layer below the internetwork-layer
```

```
-- of a network interface is considered to be an interface.

ifTable OBJECT-TYPE
  SYNTAX        SEQUENCE OF IfEntry
  MAX-ACCESS  not-accessible
  STATUS        current
  DESCRIPTION
      "A list of interface entries. The number of entries is
      given by the value of ifNumber."
  ::= { interfaces 2 }

ifEntry OBJECT-TYPE
  SYNTAX        IfEntry
  MAX-ACCESS  not-accessible
  STATUS        current
  DESCRIPTION
      "An entry containing management information applicable to a
      particular interface."
  INDEX   { ifIndex }
  ::= { ifTable 1 }

IfEntry ::=
  SEQUENCE {
    ifIndex             InterfaceIndex,
    ifDescr             DisplayString,
    ifType              IANAifType,
    ifMtu               Integer32,
    ifSpeed             Gauge32,
    ifPhysAddress       PhysAddress,
    ifAdminStatus       INTEGER,
    ifOperStatus        INTEGER,
    ifLastChange        TimeTicks,
    ifInOctets          Counter32,
    ifInUcastPkts       Counter32,
    ifInNUcastPkts      Counter32, -- deprecated
    ifInDiscards        Counter32,
    ifInErrors          Counter32,
    ifInUnknownProtos   Counter32,
    ifOutOctets         Counter32,
    ifOutUcastPkts      Counter32,
    ifOutNUcastPkts     Counter32, -- deprecated
    ifOutDiscards       Counter32,
    ifOutErrors         Counter32,
    ifOutQLen       Gauge32,   -- deprecated
    ifSpecific      OBJECT IDENTIFIER -- deprecated
  }

ifIndex OBJECT-TYPE
  SYNTAX        InterfaceIndex
  MAX-ACCESS  read-only
  STATUS        current
  DESCRIPTION
      "A unique value, greater than zero, for each interface. It
      is recommended that values are assigned contiguously
```

```
            starting from 1. The value for each interface sub-layer
            must remain constant at least from one re-initialization of
            the entity's network management system to the next re-
            initialization."
        ::= { ifEntry 1 }

    ifDescr OBJECT-TYPE
        SYNTAX       DisplayString (SIZE (0..255))
        MAX-ACCESS   read-only
        STATUS       current
        DESCRIPTION
            "A textual string containing information about the
            interface. This string should include the name of the
            manufacturer, the product name and the version of the
            interface hardware/software."
        ::= { ifEntry 2 }

    ifType OBJECT-TYPE
        SYNTAX       IANAifType
        MAX-ACCESS   read-only
        STATUS       current
        DESCRIPTION
            "The type of interface. Additional values for ifType are
            assigned by the Internet Assigned Numbers Authority (IANA),
            through updating the syntax of the IANAifType textual
            convention."
        ::= { ifEntry 3 }

    ifMtu OBJECT-TYPE
        SYNTAX       Integer32
        MAX-ACCESS   read-only
        STATUS       current
        DESCRIPTION
            "The size of the largest packet which can be sent/received
            on the interface, specified in octets. For interfaces that
            are used for transmitting network datagrams, this is the
            size of the largest network datagram that can be sent on
            the interface."
        ::= { ifEntry 4 }

    ifSpeed OBJECT-TYPE
        SYNTAX       Gauge32
        MAX-ACCESS   read-only
        STATUS       current
        DESCRIPTION
            "An estimate of the interface's current bandwidth in bits
            per second. For interfaces which do not vary in bandwidth
            or for those where no accurate estimation can be made, this
            object should contain the nominal bandwidth. If the
            bandwidth of the interface is greater than the maximum
            value reportable by this object then this object should
            report its maximum value (4,294,967,295) and ifHighSpeed
            must be used to report the interace's speed. For a sub-
            layer which has no concept of bandwidth, this object should
```

```
        be zero."
    ::= { ifEntry 5 }


ifPhysAddress OBJECT-TYPE
   SYNTAX        PhysAddress
   MAX-ACCESS    read-only
   STATUS        current
   DESCRIPTION
       "The interface's address at its protocol sub-layer. For
       example, for an 802.x interface, this object normally
       contains a MAC address. The interface's media-specific MIB
       must define the bit and byte ordering and the format of the
       value of this object. For interfaces which do not have such
       an address (e.g., a serial line), this object should
       contain an octet string of zero length."
    ::= { ifEntry 6 }


ifAdminStatus OBJECT-TYPE
   SYNTAX INTEGER {
           up(1),      -- ready to pass packets
           down(2),
           testing(3)  -- in some test mode
       }
   MAX-ACCESS  read-write
   STATUS      current
   DESCRIPTION
       "The desired state of the interface. The testing(3) state
       indicates that no operational packets can be passed. When a
       managed system initializes, all interfaces start with
       ifAdminStatus in the down(2) state. As a result of either
       explicit management action or per configuration information
       retained by the managed system, ifAdminStatus is then
       changed to either the up(1) or testing(3) states (or
       remains in the down(2) state)."
    ::= { ifEntry 7 }


ifOperStatus OBJECT-TYPE
   SYNTAX INTEGER {
           up(1),           -- ready to pass packets
           down(2),
           testing(3),   -- in some test mode
           unknown(4),   -- status can not be determined
                         -- for some reason.
           dormant(5),
           notPresent(6),     -- some component is missing
           lowerLayerDown(7) -- down due to state of
                             -- lower-layer interface(s)
       }
   MAX-ACCESS read-only
   STATUS      current
   DESCRIPTION
       "The current operational state of the interface. The
       testing(3) state indicates that no operational packets can
       be passed. If ifAdminStatus is down(2) then ifOperStatus
```

should be down(2). If ifAdminStatus is changed to up(1)
then ifOperStatus should change to up(1) if the interface
is ready to transmit and receive network traffic; it should
change to dormant(5) if the interface is waiting for
external actions (such as a serial line waiting for an
incoming connection); it should remain in the down(2) state
if and only if there is a fault that prevents it from going
to the up(1) state; it should remain in the notPresent(6)
state if the interface has missing (typically, hardware)
components."
::= { ifEntry 8 }

ifLastChange OBJECT-TYPE
  SYNTAX   TimeTicks
  MAX-ACCESS read-only
  STATUS   current
  DESCRIPTION
     "The value of sysUpTime at the time the interface entered
     its current operational state. If the current state was
     entered prior to the last re-initialization of the local
     network management subsystem, then this object contains a
     zero value."
  ::= { ifEntry 9 }

ifInOctets OBJECT-TYPE
  SYNTAX   Counter32
  MAX-ACCESS read-only
  STATUS   current
  DESCRIPTION
     "The total number of octets received on the interface,
     including framing characters.

     Discontinuities in the value of this counter can occur at
     re-initialization of the management system, and at other
     times as indicated by the value of
     ifCounterDiscontinuityTime."
  ::= { ifEntry 10 }

ifInUcastPkts OBJECT-TYPE
  SYNTAX   Counter32
  MAX-ACCESS read-only
  STATUS   current
  DESCRIPTION
     "The number of packets, delivered by this sub-layer to a
     higher (sub-)layer, which were not addressed to a multicast
     or broadcast address at this sub-layer.

     Discontinuities in the value of this counter can occur at
     re-initialization of the management system, and at other
     times as indicated by the value of
     ifCounterDiscontinuityTime."
  ::= { ifEntry 11 }

ifInNUcastPkts OBJECT-TYPE

```
SYNTAX      Counter32
MAX-ACCESS  read-only
STATUS      deprecated
DESCRIPTION
    "The number of packets, delivered by this sub-layer to a
    higher (sub-)layer, which were addressed to a multicast or
    broadcast address at this sub-layer.

    Discontinuities in the value of this counter can occur at
    re-initialization of the management system, and at other
    times as indicated by the value of
    ifCounterDiscontinuityTime.

    This object is deprecated in favour of ifInMulticastPkts
    and ifInBroadcastPkts."
::= { ifEntry 12 }

ifInDiscards OBJECT-TYPE
    SYNTAX      Counter32
    MAX-ACCESS  read-only
    STATUS      current
DESCRIPTION
    "The number of inbound packets which were chosen to be
    discarded even though no errors had been detected to
    prevent their being deliverable to a higher-layer protocol.
    One possible reason for discarding such a packet could be
    to free up buffer space.

    Discontinuities in the value of this counter can occur at
    re-initialization of the management system, and at other
    times as indicated by the value of them
    from being deliverable to a higher-layer protocol.

    Discontinuities in the value of this counter can occur at
    re-initialization of the management system, and at other
    times as indicated by the value of
    ifCounterDiscontinuityTime."
::= { ifEntry 13 }

ifInErrors OBJECT-TYPE
    SYNTAX      Counter32
    MAX-ACCESS  read-only
    STATUS      current
DESCRIPTION
    "For packet-oriented interfaces, the number of inbound
    packets that contained errors preventing them from being
    deliverable to a higher-layer protocol. For character-
    oriented or fixed-length interfaces, the number of inbound
    transmission units that contained errors preventing them
    from being deliverable to a higher-layer protocol.

    Discontinuities in the value of this counter can occur at
    re-initialization of the management system, and at other
    times as indicated by the value of
```

```
                    ifCounterDiscontinuityTime."
           ::= { ifEntry 14 }


  ifInUnknownProtos OBJECT-TYPE
      SYNTAX         Counter32
      MAX-ACCESS     read-only
      STATUS         current
      DESCRIPTION
          "For packet-oriented interfaces, the number of packets
          received via the interface which were discarded because of
          an unknown or unsupported protocol. For character-oriented
          or fixed-length interfaces that support protocol
          multiplexing the number of transmission units received via
          the interface which were discarded because of an unknown or
          unsupported protocol. For any interface that does not
          support protocol multiplexing, this counter will always be
          0.

          Discontinuities in the value of this counter can occur at
          re-initialization of the management system, and at other
          times as indicated by the value of
          ifCounterDiscontinuityTime."
      ::= { ifEntry 15 }


  ifOutOctets OBJECT-TYPE
      SYNTAX         Counter32
      MAX-ACCESS     read-only
      STATUS         current
      DESCRIPTION
          "The total number of octets transmitted out of the
          interface, including framing characters.

          Discontinuities in the value of this counter can occur at
          re-initialization of the management system, and at other
          times as indicated by the value of
          ifCounterDiscontinuityTime."
      ::= { ifEntry 16 }


  ifOutUcastPkts OBJECT-TYPE
      SYNTAX         Counter32
      MAX-ACCESS     read-only
      STATUS         current
      DESCRIPTION
          "The total number of packets that higher-level protocols
          requested be transmitted, and which were not addressed to a
          multicast or broadcast address at this sub-layer, including
          those that were discarded or not sent.

          Discontinuities in the value of this counter can occur at
          re-initialization of the management system, and at other
          times as indicated by the value of
          ifCounterDiscontinuityTime."
      ::= { ifEntry 17 }
```

```
ifOutNUcastPkts OBJECT-TYPE
   SYNTAX       Counter32
   MAX-ACCESS   read-only
   STATUS       deprecated
   DESCRIPTION
      "The total number of packets that higher-level protocols
      requested be transmitted, and which were addressed to a
      multicast or broadcast address at this sub-layer, including
      those that were discarded or not sent.

      Discontinuities in the value of this counter can occur at

      re-initialization of the management system, and at other
      times as indicated by the value of
      ifCounterDiscontinuityTime.

      This object is deprecated in favour of ifOutMulticastPkts
      and ifOutBroadcastPkts."
   ::= { ifEntry 18 }

ifOutDiscards OBJECT-TYPE
   SYNTAX       Counter32
   MAX-ACCESS   read-only
   STATUS       current
   DESCRIPTION
      "The number of outbound packets which were chosen to be
      discarded even though no errors had been detected to
      prevent their being transmitted. One possible reason for
      discarding such a packet could be to free up buffer space.

      Discontinuities in the value of this counter can occur at
      re-initialization of the management system, and at other
      times as indicated by the value of
      ifCounterDiscontinuityTime."
   ::= { ifEntry 19 }

ifOutErrors OBJECT-TYPE
   SYNTAX       Counter32
   MAX-ACCESS   read-only
   STATUS       current
   DESCRIPTION
      "For packet-oriented interfaces, the number of outbound
      packets that could not be transmitted because of errors.
      For character-oriented or fixed-length interfaces, the
      number of outbound transmission units that could not be
      transmitted because of errors.

      Discontinuities in the value of this counter can occur at
      re-initialization of the management system, and at other
      times as indicated by the value of
      ifCounterDiscontinuityTime."
   ::= { ifEntry 20 }
```

```
ifOutQLen OBJECT-TYPE
   SYNTAX       Gauge32
   MAX-ACCESS   read-only
   STATUS       deprecated
   DESCRIPTION
       "The length of the output packet queue (in packets)."
   ::= { ifEntry 21 }

ifSpecific OBJECT-TYPE
   SYNTAX        OBJECT IDENTIFIER
   MAX-ACCESS   read-only
   STATUS       deprecated
   DESCRIPTION
       "A reference to MIB definitions specific to the particular
       media being used to realize the interface. It is
       recommended that this value point to an instance of a MIB
       object in the media-specific MIB, i.e., that this object
       have the semantics associated with the InstancePointer
       textual convention defined in RFC 2579. In fact, it is
       recommended that the media-specific MIB specify what value
       ifSpecific should/can take for values of ifType. If no MIB
       definitions specific to the particular media are available,
       the value should be set to the OBJECT IDENTIFIER { 0 0 }."
   ::= { ifEntry 22 }
```

Now that we have three related tables under our belts, it should be easier to see how a single name space helps us associate the information found in different tables. By the same token, when all kinds of management information (fault, configuration, etc.) are incorporated in a single name space, it becomes easier to see relationships between these data elements. Figure 6.3 is a graphic representation of the description of the three related tables we are discussing.

Although the example above attests to the possibility of relating information from fault, configuration, and other systems in one or more MIB Modules, it is far from being an ideal example. The data is not as well organized as it could be—specifically, the IETF does not impose any organization of data in the assignment of object identifiers for different MIB Modules. The base object identifier (OID) for each MIB Module is simply assigned with the next available number. As we will see later, structure can instead be created with object identifiers and used to great advantage.

Perhaps the best example of different name spaces in the management software environment is the one between SNMP and vendor CLIs. Because CLIs and other access methods always persist to some degree, I argue the need for a single common name space onto which we can map other methods. The practice of basing management objects on the SNMP

Figure 6.3
Integrated
information in a
common name
space.

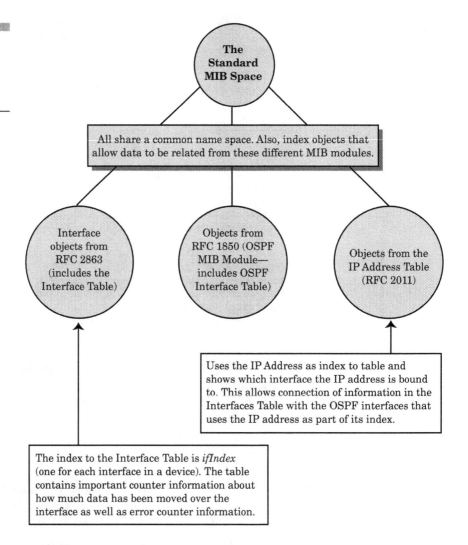

name space and mapping other name spaces to it is increasingly popular in commercial products. (This concept is explored in more detail in Chapter 8.)

In the absence of a common name space, the cost of management software for managed devices increases, while its effectiveness declines. The primary reason is because "glue" software must be written to translate or map between different name spaces. Translation takes several forms:

▪ Translation between individual management objects at both a semantic level and a simple naming level. The meaning of the SNMP object

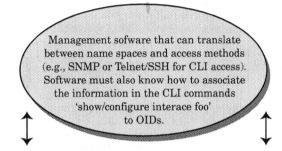

Figure 6.4
Using SNMP for a common name space.

Objects in the MIB name space
1.3.6.1.... such as:
• ifInOctets
• ifOperStatus
• ifAddEntAddr
•ospfIfDesignatedRouter

Objects in the command line interface
name space which is always vendor
specific such as:
• interface counters
• configure interface (parameters)

SNMP
Name Space
(The MIB)

CLI
Name Space
(The Command
Hierarchy)

Instrumentation inside the managed device that contains the management
information such as that found on an interface card for how many packets
have been received or inside the operation system for how much memory
or CPU a process such as a Web server is using.

ifInOctets is in the DESCRIPTION clause of the MIB Object that
defines ifInOctets. Vendors may have implemented many different
CLI commands, each of which could have the same or nearly the
same meaning as ifInOctets. As a result, we might get one value
when retrieving ifInOctets using SNMP, and a different value when
we use the CLI on that same device to retrieve what we think is the
same information as obtained by ifInOctets. Similarly, this command
line version is likely to differ from one vendor to another, thus adding
further confusion.

▧ Translation between object instance identification. Another problem
with multiple name spaces is that the method of instance identification
usually varies. Even when numeric values are used to identify an
object instance, there is no guarantee that the software realizing

Figure 6.5

Differences in object meanings and instance identification with different name spaces.

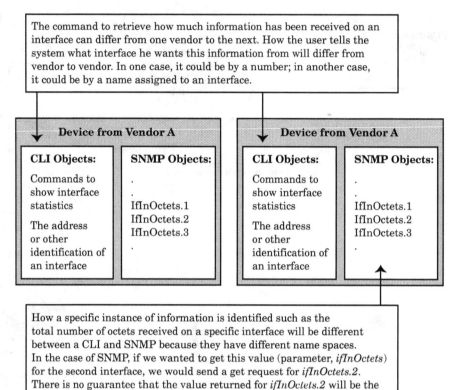

The command to retrieve how much information has been received on an interface can differ from one vendor to the next. How the user tells the system what interface he wants this information from will differ from vendor to vendor. In one case, it could be by a number; in another case, it could be by a name assigned to an interface.

Device from Vendor A		Device from Vendor A	
CLI Objects:	**SNMP Objects:**	**CLI Objects:**	**SNMP Objects:**
Commands to show interface statistics	. . IfInOctets.1 IfInOctets.2 IfInOctets.3 .	Commands to show interface statistics	. . IfInOctets.1 IfInOctets.2 IfInOctets.3 .
The address or other identification of an interface		The address or other identification of an interface	

How a specific instance of information is identified such as the total number of octets received on a specific interface will be different between a CLI and SNMP because they have different name spaces. In the case of SNMP, if we wanted to get this value (parameter, *ifInOctets*) for the second interface, we would send a get request for *ifInOctets.2*. There is no guarantee that the value returned for *ifInOctets.2* will be the same as the request to the CLI because of how we name these interfaces and implementation differences.

objects in each name space will use the same methods to assign and increment the index values as its counterparts do in other name spaces.

▪ Translation between the semantics of object groupings (e.g., a row in a table) and a series of related CLI commands. In SNMP, objects can be related in MIB Tables, in MIB Groups, and between tables contained in different MIB Modules. CLIs go about relating parameters somewhat differently. One common approach is to put a series of related parameters together on a command line and stipulate that they all must be present for the command to be valid. In other cases, some command parameters are optional. In SNMP, whether an object can or must be present is a part of its definition.* In either case, mechanisms exist for providing information to the user or application about invalid object combinations. Environments where multiple name spaces are used rely on management software to know how to process each different error return mechanism in the system.

Translation between different name spaces on a managed device is not mandatory. An alternative is to implement two completely different access methods and their associated communications infrastructures, along with the upticks in memory, CPU, and engineering resources needed to implement these access methods. We will defer discussion of implementation alternatives to the sections on designing management software for managed devices.

Management application software suffers from differences in name spaces. In addition to the considerations above, the management software must create associations between different management objects and apply value to those associations. For example, if an interface has been configured to support an important customer, and a notification indicative of a failure is received, the software may react very differently from the way it would if that interface had been configured for low-priority traffic to another customer. These associations are much easier when all the objects are in the same name space. In most systems today, the configuration information is sent to the managed device via CLI commands or file transfer. An indication of an interface failure is most likely to come via SNMP. As a result, the management station must make the association between the configured status of the interface and the failure. A harder task with these two separate name spaces.

Different name spaces often imply different frameworks: CLI configuration systems and SNMP-based data collection engines are two examples. Where multiple frameworks are used, configuration data must be sent by one software system to the managed devices while data is collected from these managed devices with a different (probably SNMP-based) system. The configuration information must get from the configuration tools to the data-collection software so that it knows what data to collect and where to get it. This process can be manual or the result of custom integration software. In either case, the introduced complexity raises costs. Another hidden cost to different name spaces is that many data col-

---

*There are several facilities in SNMP for the specification of whether an object can be optional or must be present. The STATUS clause [RFC2578] indicates whether the object is obsolete (no longer used), deprecated (it is obsolete but can be used), or current. In the case of an object that is defined as part of a conceptual row in a table, the DEFINITION clauses of the object and related RowStatus objects in that row indicate objects required for the row to be considered valid. See more on the RowStatus textual convention in [RFC2579]. Lastly Conformance Statements for SMIv2 describe ways that MIB writers can group objects together and indicate under what circumstances they are required. The AGENT-CAPABILITIES macro, also described in [RFC2580], can be used by vendors to express the level of conformance their particular implementation has to a MIB Module. Unfortunately, these macros are seldom provided in such way that a management system can read them and make adjustments based on what is in the capabilities macro.

lection tools attempt to verify what to collect from the managed devices themselves: in SNMP-based data-collection systems, verification is often accomplished via retrieval of OIDs in different areas of the SNMP MIB. Although this method is useful, it does not always convey what the configuration "intent" is. For example, it may not relay that data doesn't have to be collected from every interface in a device, if some interfaces won't be used in a way that requires detailed monitoring. Plainly, omitting that information about which interfaces need not have data collected from them increases the load on the network and the managed devices.

Although the foregoing discussion has focused on SNMP and the CLI as the two different name space examples, similar problems exist with any two systems that might be used in conjunction. The degree of difficulty for reconciling different name spaces in the management software, whether inside a device or in the management applications themselves, will be a function of *how different* the name spaces are. There is a great likelihood that the market will not converge on a single name space in the foreseeable future,* so the best we can do is minimize these difficulties and build management software that does an efficient job of compensating for name space differences.

Relatively Simple Protocol Operations

Albert Einstein said "Things should be made as simple as possible, no simpler." Our goal is to ensure that the framework for our management system contains all the low-level features that are needed, but none that may tend to increase implementation complexity and resource usage. There is room for reasonable people to differ on how many operations are enough. SNMP has very few protocol operations, making it easier to implement. Yet, some argue that SNMP is so stripped down that management applications become more complex to build and less efficient to operate.

A natural layering should take place in a management environment: the complexity of sophisticated management operations should be isolated from the low-level protocol, and the upper-layer management application level should be responsible for complex transaction state management and other sophisticated functions. Some believe that the protocol should also embed some level of transaction control. In practice, this isn't feasible, because multiple transactions can only be understood by the

*A rapid convergence is not likely for many reasons. In short, many independent vendor interests are pressing for different technologies, and standards organizations continue to generate new alternatives.

management application software and the software that executes inside each managed element. Nonetheless it has some prima facie appeal. In addition to their natural tendency to look to the protocol level for relief of complexity, some people find it hard to distinguish data integrity from transaction control. *Data integrity* is the assurance that what is delivered is what was sent and, if it is not, that the error is detected.* *Transaction control* refers to the process whereby a unit of work, such as changing all the configuration parameters on an interface, is identified and managed to completion. Transaction control is usually specific to the context of what one is doing, like configuring an interface, an entire router, or a Web server. Data integrity is a simpler principle—we just want to make sure that what was transmitted is what was received.

Proponents of TCP-based transports are among those who often blur the distinction between reliability and integrity. They incorrectly believe that having a reliable transport—which TCP is—gives them better transaction control. If they persist in the attempt to build transaction control into the protocol, the protocol becomes more complex and the original problem goes unsolved since real transaction control can only be effectively managed at the application level.†

Reasonable Efficiency

Keep these concepts firmly in mind:

* Properly understood, efficiency is a property of the entire system rather than of any one aspect.

---

*Those familiar with the term data integrity when used in the context of security concerns will expand this definition a bit. That expansion does not negate this definition. [RFC2574] User-Based Security Model (USM) for Version 3 of the Simple Network Management Protocol (SNMPv3) states: "Data integrity is the provision of the property that data has not been altered or destroyed in an unauthorized manner, nor have data sequences been altered to an extent greater than can occur non-maliciously." The concern here is the same: that what we sent is what was received. Here, the focus is to ensure that nothing malicious took place. In our working definition, we just want to make sure that the application or the network did not mess things up. In the end, we want the more expansive security-conscious definition.

†As with all things, this is a matter of degree. At the protocol level, there must be a way for two systems to know that at some automatic level, the sender has finished. The point being made is not to go overboard and overcomplicate the protocol. Some say that SNMP has gone too far the other way and does not have sufficiently sophisticated mechanisms for the transmission of configuration information and as a result, it is too hard to use for this purpose. There is validity to this concern. Discussions are underway as to how to improve SNMP in the regard.

▨ When evaluating the efficiency of a framework or system, remember
that a good system can be incorrectly optimized.

A classic example of a suboptimal set of trade-offs can be traced to the
early days of SNMP, when some of the first decisions about which MIB
Objects to create and which to omit were made. In the majority of cases,
when designing MIB Modules, the designers made the agent as simple as
possible and pushed complexity to the management station software. It
sounded right, but it produced at least two problems. First, extra protocol
operations were made on the managed device. This larger number of
operations was potentially detrimental to performance, more so if the
network was a low-speed one than if "smarter" objects had been defined.
Second, management applications had to piece together information that
they should have been able to retrieve intact from the managed systems.

Software designers classically posit that if a piece of information can be
synthesized from several data elements in a managed system, it's wrong
to require that the managed system contain the synthesized data as well.
Instead, the manager should retrieve all relevant data elements and then
perform the synthesis. This belief is not unreasonable in itself, but it has
been overstepped in practice. Care must be exercised not to apply the
rules too rigidly, because when they are, an otherwise efficient framework
can become quite the opposite. Here are some considerations to factor in
when creating managed objects and the software to realize them:

▨ **Protocol operations**—How effective are they? Is each one a major
consumer of network or equipment resources? Can they be optimized?
For example, if security figures in the framework, can it be selectively
used? Encryption consumes such a large amount of computing
resources that it should be applied only when operational considerations demand its use. Using SNMP you can select by user, operation,
and data element. You can even select the type of security to be
applied, down to the instance level.* Thus, you can apply the security
method you need only when you need it. Other approaches—say, a CLI
over TLS or solutions using IPSEC—have far less flexibility. In particular, these other approaches do not have the access control granularity
needed for the content of the management payload they carry.

---

*As a specific example, imagine a device with many network interfaces. In the case of a
system that was shared by many, it may be desirable to show instances of certain interface
counters only to specific users. For example, we might only want to show the instance of
ifInOctets on interface 1 to users of that interface, while an instance of *ifInOctets* on interface 2 might be available to others.

- **Connection methods to devices**—A great deal has been written about connection versus connectionless protocols. There are benefits to both approaches depending on circumstances. Quick data retrieval or reasonably sized configuration commands sent to a large number of devices are better handled with UDP, a connectionless protocol. UDP minimizes the overhead for each connection to the managed device. On the other hand, where very large amounts of data must continuously flow from one device to another, a connection-oriented approach is advisable. For management software, the question is how well a management system would scale, considering the large volume of continuous data transmission between each device and its management station. An early review of the pros and cons associated with TCP and UDP as transports for SNMP was documented in [RFC1270] in 1991. It is still worth reading.

- **Network utilization**—Because network speeds have increased so dramatically in the past few years, some people are persuaded that the upward trend is likely to continue. Although this may true, it is less true at the edges of a network.* Network utilization, like other aspects of efficiency, is a function not only of the technology's capabilities, but of how they are applied.

- **Memory/CPU utilization**—As with network speed, memory and CPU capacity have been increasing at a rapid rate in network devices. Yet for many products, especially at the low end of the cost spectrum, it remains critical to use these resources efficiently.

- **Implementation ease**—We've argued that implementation ease is a function of the complexity of the technology, but it also turns on the availability of public domain implementations, reference materials, and other issues. Another determinant is how understandable the technology is, and how well it is understood by the people influencing it at any given moment.

- **Management applications**—Without effective management applications, an investment in management infrastructure inside managed devices is worthless. Unless management applications can be cost effectively created and deployed, we get little out of our investment in the managed device.† Ongoing tension exists between the complexity

---

*Network utilization is also a concern to enterprises that run their own private networks. They pay for all their lines, so getting the most out of each is a real concern. Minimizing the amount of management traffic on relatively slow links can reduce the need to upgrade to more expensive links.

†The ubiquity of effective SNMP-based data collection, fault management, reporting, and root-cause analysis tools shows the power of having a common infrastructure in managed devices (at least common enough). One application can be written to perform these functions for a wide variety of vendors. We have yet to reach this state in the configuration realm in IP environments using any technology.

of the software executing in a managed element and the complexity it demands of the coordinating management application. Too far in either direction results in an inefficient system.

Extensibility

Whichever framework you choose, it should be designed with a range of extensibility in mind. Types of extensibility include:

- Facility for different vendors to define their own management objects without stepping on each other's name space.
- Ability to add new protocol operations as needed.
- Ability to add new data types and language primitives as needed.
- Ability to add new transport mappings as needed.
- Ability to support and use different security mechanisms, such as data encryption, without changing the base protocols.

Integrating Different Types of Management Data

Closely related to extensibility is the framework's ability to represent and process all types of management data: fault, configuration, accounting, performance, and security. Some systems represent configuration information very well, while others aren't sufficiently optimized for that purpose. Similarly, some systems manage configuration information quite well, but are wholly inadequate to when it comes to fault, performance, security, or accounting functions. A framework not only has to represent management information competently, but it also has to collect, process, and distribute the information in an appropriate way. For configuration information, the priority is concise transmission and efficiency of operation. CLIs and configuration files have been the preferred vehicles for configuration, but as systems have become larger and more elaborate, the size of the files and number of commands have begun to cause significant scale concerns. CLIs and files that represent the CLI commands have not yet been extended to meet this challenge. On the other hand, SNMPCONF is working to do so for SNMP-based systems.*

---

*The cost of this efficiency for SNMPCONF is the requirement that more intelligence be distributed to the managed devices. If objections to this distribution of intelligence can be overcome, then we will have a system that includes configuration, fault, performance, security and accounting data, since SNMP has a long history in these other areas.

A similar situation exists for data collection. Such a framework must not only be able to adequately represent a data collection object—such as an interface counter—but it must also collect such data efficiently. Connection-oriented frameworks have a historic difficulty in this area.

The ISO Framework that includes the Common Management Information protocol [CMIP] is the only other framework that comes close to SNMP in terms of completeness in all of these areas,* except that it lacks a security framework similar to the one provided by SNMP. It has not been well accepted in the IP environment however.

Robustness

A number of important features combine to make the SNMP framework robust.

Registered Entities

Many thousands[†] of entities have registered with the Internet Assigned Numbers Authority [IANA] to create their own extensions to the SNMP MIB. This is both one of the greatest successes and greatest challenges for SNMP. Vendors can extend the management environment to suit the needs of their products, thus ensuring their market differentiation. Unfortunately, many vendors implement objects in their portion of the name space that are either duplicate or gratuitously different from objects already defined in the standard portion of the name space. The problem of extraneous objects is not related to any aspect of the SNMP technology. It is a function of vendor indifference and lack of customer pressure on vendors to "do the right thing."

---

*In many ways, the ISO management framework is more complete than SNMP. It has a richer set of functions than SNMP. However, it appears as if the complexity trade-off made in the ISO domain went too far in the direction of functional complexity, given the very poor adoption rate (in the IP environment).

†As of November 2001, there were over 11,000 listings of vendors with private enterprise numbers. To see a full listing go to http://www.isi.edu/in-notes/iana/assignments/enter-prise-numbers.

Data Collection

The SNMP framework also works well for data collection. When the protocol evolved from version 1 to version 2, several new operations were added to meet the demand for improved efficiency for data retrieval and to provide confirmation of asynchronous notifications [RFC1905]. At the time of this writing, a working group is dedicated to investigating further enhancements to the protocol as user needs continue to evolve [EOS].

Over time, SNMP has added new data types and features to the language it uses for object definition [RFC2578]. A new working group is investigating further evolution of this language [SMIng], and the framework is easily keeping pace with those introduced more recently.

Transport

The SNMP has supported many transport mappings for some time. The fact that UDP remains the dominant and sole required mapping suggests that, for the most part, this transport has done the job. Proposals for standard support for TCP are on the table and may get adopted, but my suspicion, based on experience with connection-based management systems, is that TCP will be used only infrequently even then. Maintaining an open TCP connection to every system it manages demands more of a management station that UDP does. That is not to say, however, that we should not invest in the work to make a TCP-based alternative available. The availability as a TCP-based alternative along with some additional enhancements proposed by the EOS and SMIng working groups will greatly improve the utility of SNMP, particularly for configuration.

Security

The SNMP framework version 3 has shown how it accommodates a variety of security mechanisms even though one is mandatory. As with its primary reliance on UDP for a transport, SNMP's primary reliance on a single recommended set of security mechanisms is intended to support interoperability while preserving extension [RFC2574, RFC2575].

A Formalized Language for the Representation of the Data

A number of alternative technologies do one or more of the SNMP functions "better" than SNMP. In fact, the recent history of management software is cluttered with technologies that were going to take over management because of one feature or another (e.g., Web-Based Management, the ISO CMIP Protocol, the Open Software Foundation's Distributed Management Environment, XML, etc.). We have learned from this endless progression of technologies that it is not any individual feature that matters when creating a framework (indeed many of the technologies have outstanding qualities). What is much more important is the overall effectiveness of all components of a framework working together, along with the trade-offs made to get them to work together as a whole. So, while this discussion may seem to some skeptics like the apologist's guide to SNMP, the intent is more pragmatic: to illustrate important features required of any technology if it is to be seriously considered as a framework for a comprehensive service management system. An analysis based on other criteria might yield quite different results. The next sections illustrate some shortcomings of the venerable SNMP framework and then argue why, even in the face of those shortcomings, SNMP is currently the best alternative for a service management system based on a single framework.

Drawbacks to the SNMP Framework

Within the context of a complete service management system, the features described in the previous section are all very important, but collectively they're focused on how to name, secure, and move the data. That's a good place to start, but not to stop. What else belongs in a full service management system? The standards are silent on implementation details in managed elements and management applications. Details are missing on:

* how to integrate SNMP in managed devices with other access mechanisms, and how to structure the management data;

- how to design management applications for the effective processing of management data received from managed devices;
- what managements applications can and should do before sending notifications or set commands to managed devices;
- human interface design; and
- implementation approaches.

One could argue that it is outside of the purview of a standards body, (the IETF in the case of SNMP), to specify user interface and internal design approaches. The purpose of the framework is, after all, to provide some basic building blocks to which we can add to create a full solution. In many respects, I agree with the view that these areas are outside the legitimate domain of standards bodies, at least for the present. This leaves equipment vendors, application developers, and network operators on their own when it comes to filling in these details. Much of the rest of this book is a set of suggestions for these areas.

The IETF has published informational [RFC3139], Requirements for Configuration Management of IP-based Networks, describing some general configuration management requirements; it is useful background reading. RFC 3139 is an enumeration of requirements for configuration management that designers may wish to examine before embarking on a system design.

Why Using a Single Framework Is Beneficial

Throughout this section I have espoused the desirability of a single framework from many perspectives. A single framework is not only technically beneficial; it is just as important from a financial perspective. Here is a good place to review the reasons to use a single framework, instead of a system built on several frameworks:

- **Protocol operations**—Semantic differences are eliminated. In a single-framework environment there are no mappings to perform and no possibility for semantic differences between them. This coherent environment also makes it somewhat easier to map to databases found in the management applications, where that makes sense. For example, if an application is performing a transaction across many network devices, the application does not have to be concerned with semantic differences between one protocol and the next when only one protocol is used. The application is leaner, because there is only one protocol to implement.

* **Security infrastructure**—Coordination and integration are eliminated. Wherever you find multiple security infrastructures, they are either "ships in the night," or they have been laboriously reconciled. In the case of the ships, each security system must be configured separately. Worse yet, humans have to coordinate access privileges between the different security systems so that access can't be granted inadvertently. In the simplest example, expect to configure both SNMP and CLI access security information on each device. Supposing you take an integrated approach to the presence of multiple frameworks, each with its own security infrastructure, expect to write specialized code for mapping the different infrastructures. Unfortunately, it's rare that the levels of access and control map easily between frameworks, thus making the task of developing the security mapping software nontrivial.

* **Object name space**—Incompatibilities and mapping difficulties are eliminated. One-to-one mappings of objects in different name spaces are often not obvious, and sometimes not even available. This problem makes the integrated management of an environment with mixed name spaces problematic.

* **Management station software**—Efficiency is improved. Management systems must devote code, memory and CPU cycles to translation from one framework to another. This is true of name spaces, security systems, protocols, and other items. While it is possible to develop code that can perform many of these functions, it will be code that does not really facilitate management. What is derisively termed "glue code" simply sticks things together so that management functions can proceed. Glue code is not a neutral issue—it is one of the reasons integrated management systems do not exist today. Instead, we have *stovepipe solutions* within a single framework. They can provide some real value without incurring the cost of integration, but they leave the integration work to the unhappy customer.

* **Accounting and billing**—In some cases, it is not possible to generate accurate bills without an integrated system, because the information needed to do it (configuration, state changes during the billing period, over- and underutilization based on contracts, etc.) is spread

across the system. This is not to imply that an acceptable integrated system must be able to implement full billing and accounting systems. It is, however, non-negotiable that they can capture all of the needed information to pass on to billing and accounting.

- **The cost of management**—Management for each device, configuration, and deployment of management software is cheaper. The implementation of multiple frameworks costs a lot in money and hardware resources. A single framework reduces the number of protocol stacks and mapping between frameworks, their name spaces, and security systems in managed devices. Beyond that, to the degree we wish to integrate separate frameworks, information from one must be mapped to the other (in the managed device, management application, or both). All of this typically raises deployment costs significantly, and typically we do not get the level of integration desired.

- **Multivendor management software**—Integrating management software from different vendors is easier. If each vendor uses a common framework, then it is more likely that they can integrate at the protocol or data levels.

- **Fault and SLA management**—Keeping promises to customers is easier. Because faults and service-level agreements are understood in the context of configuration information and the promised levels of service, a system that does not integrate all these data cannot hope to provide the management facilities to control services. At a simpler level, without knowing how a device is configured, you cannot adequately evaluate and respond to failures and performance bottlenecks.

- **Transacting across name spaces**—A set of difficult problems associated with transactions in different name spaces can be sidestepped.

- **Cost of configuration and operation of several different suites of management software**—The cost of configuring and operating software is reduced when we have an integrated system. With different applications, in which configuration (or other) information about services and devices in the network has to be replicated (such as in configuration tools and data collection systems) due to uncommon infrastructures for each of the applications, we have increased operational costs. This is not necessary when we have an integrated system built around a single management framework.

So Why Use SNMP as a Basis for a Service Management System?

First, do the math. If you accept the principle that a single framework is important and that an effective framework for service management includes most, if not all, of the features we have discussed, then your choices are two: SNMP and the OSI Framework. We have already downgraded the OSI approach because it never really caught on in IP environments, so we are left with SNMP.

Some compelling reasons for using SNMP as our framework exist, beyond the fact that nothing else seems to meet the requirements:*

* **Applications**—SNMP offers a huge installed base of management applications used primarily for data collection and fault management. From these, we've learned a great deal about SNMP-based applications and how to integrate them with database technology and other important aspects of a management system. The industry sorely needs applications that can perform configuration and security management functions, but so far very few exist. A knowledge base like SNMP's can be leveraged, and should prove invaluable as we develop more complete solutions.
* **Standard measurement objects**—For an effective service management system, one needs not only to configure the services but also to collect usage and state information. Collectively, members of the industry have a huge investment in both standard and vendor-specific objects to support service verification, fault, and accounting. That investment represents gains individuals can leverage by adding configuration information, rather than replicating it in other frameworks or attempting to glue it to partial configuration-only–based systems.
* **Installed base of agents and instrumentation**—Here the investment is not just in the management objects themselves; it is also in deployed code in network elements of all kinds, from small desktops, to servers, to cable modems, and on up to large routers at the core of the Internet.

---

*For those who claim that the flaws they see in SNMP disqualify it as our choice, few would have another alternative. Until something better comes along, SNMP provides us a way to build the systems we need. Real-world experience will determine how widely and well this approach can work. I also want to emphasize once more that the general principles we have discussed to this point are at least as important as the application of SNMP to the realization of those principles in a service management system.

- **Software development familiarity with the technology**—The level of sophistication in many organizations that develop SNMP-based technology for deployment in networks is not always highly advanced. Even so, there is a critical mass of engineering understanding to be preserved and leveraged. The only reason these engineers lack more highly advanced skills has little to do with SNMP, but rather is because corporations have not placed much emphasis on management technologies in general.

- **Proven in network operations**—SNMP is the most widely deployed technology devoted to management in IP-based networks today. We know it works, and works well, in a variety of environments. When things go wrong, we probably know why. Most network staffers don't have sophisticated enough understanding of SNMP for the development of SNMP-based technology, but nearly all have considerable background knowledge and a comfort level with how SNMP operates when deployed.

- **Flexibility**—SNMP is used in environments from small enterprise to the Internet core, in servers and desktop devices. It has proven adaptability not only to different operational environments, but also in a wide range of applications from power supply management to industrial database servers.

- **A fully fleshed out, user-based security model**—Particularly important when performing configuration and other control functions, the SNMP infrastructure supports access to information based on both users and roles. Therefore, it permits the deployment of management software that conforms to actual operational models.

- **Name space infrastructure management in place**—IANA has been performing this function for the SNMP name space since the beginning of time, and the arrangement has worked quite well. It allows extension to the name space by vendors and still guarantees uniqueness. Yes, that could be accomplished with other technologies as well, but as we have already observed, one name space is preferable.

---

*Witness the huge installed base of these types of applications. These tools are in widespread, continuous usage in almost all IP networks of any size. While they are not used widely for configuration, in part because the IETF has failed to make a critical mass of configuration objects available, experience from vendors indicates that SNMP can do the job. Some vendors have products in common use today that have the configuration management function performed effectively with SNMP through the use of private MIB objects. The cable modem industry has demonstrated that SNMP can form the foundation of an effective configuration management system based on standard MIB objects.

†Examples of evolving technologies managed by SNMP include the OSPF and BGP routing protocols.

The most compelling reason for using SNMP is that it can do the job. It has performed well in fault management and data collection for accounting and performance information for a number of years.* It has been amended to meet new needs over and over again. These changes notably include the addition of new, more-efficient protocol data units like the GetBulkRequest-PDU, a new security infrastructure, and new data types. This evolutionary approach, which SNMP has exemplified, is consistent with other protocols used in the Internet.†

From the foregoing, we can conclude that:

* SNMP requires improvement, just like any protocol serving needs that are changing. That's not worrisome, because SNMP has been successfully improved more than once, and a number of groups (EOS, SMIng, and SNMPCONF) are underway working on important enhancements.
* Evolving the Internet Standard Management Framework, rather than developing a completely new one (or worse yet, a series of unrelated partial solutions), will give us what we need faster and with less cost than the other alternatives.

With all of this said, some people will reject any system based on SNMP precisely because they believe that partial solutions or a completely new solution are the best course for the future. I suggest that they read the remaining chapters, not for the specifics of SNMP technology, but for the principles on which an effective management system can be built.

CHAPTER 7

Configuration Management Using SNMPCONF

So far we have defined network services, general requirements for a system that can manage these services, and the special problems that exist at network edges. Early chapters looked at some of the technologies that attempt to address the special problems of managing complex services in a large network environment. With this background, we created a model to use to view the problem of service management. Chapter 6 laid out the requirements for a framework of service management and why, of the choices currently available, SNMP is the optimum foundation of a complete service management system, given today's technologies.

In this and subsequent chapters we start to flesh out a complete solution using SNMP and SNMPCONF. Equal attention will be paid to managed devices and management applications, since one is of little value without the other. SNMPCONF technology is quite new. How widely it will be adopted and its effect on the market are still unknowns. I present it here as a way to build service management into network devices purely on the grounds that it meets the requirements. Should other technologies be adopted, many of the principles will be the same as those assumed by the SNMPCONF approach, because these other technologies must also reflect the principles presented in the preceding chapters, if they are to meet the demands of a service management system.

Note that the SNMPCONF technology is so new that, at the time this book was being written, the working group had not yet completed the final revision of its work. The [PM] reference is the latest available Internet draft and, as such, should be thought of as a work in progress that is likely to change. What will not change, at least for the purposes of this book, are the important points about building a service management system, which this technology illustrates. In short, the details may change, but not the essential points.

Background and Motivation

SNMPCONF resulted from of the growing conviction among SNMP workers that, to meet the management challenges of networks with more complex systems and the services that were layered on those systems, we needed a complete solution rather than a series of specialized single-purpose protocols.

A good deal of discussion took place in the IETF about the merits of various ways to meet the needs of advanced configuration management.

A number of meetings were held and Internet drafts were published presenting the alternatives. Since Internet drafts are not held in the IETF repository beyond six months after publication, a draft relevant to this discussion has been preserved at [SNMPCONF-BACKGROUND].* Many of the requirements described below were first synthesized in SNMPCONF-BACKGROUND.

Network engineers and developers wanted a system that:

* Was based on technology that was already *well understood* and *widely deployed*, so that incremental progress could be made in a short time.
* Was based on a *single* technology, so that we could integrate fault, configuration, performance, accounting, and security data more easily.
* Was built on a *complete* framework, so that in addition to moving data between management systems and devices, we also had to have a serious security system. The system also required all the other facilities associated with SNMP, such as processing the asynchronous messages sent via the TRAP or InformRequest-PDUs.
* Incorporates valuable policy-based principles of abstraction and encapsulation. One important benefit of these principles is that they allows us to send data in a more efficient manner than we currently do with SNMP- or CLI-based systems, because it is not necessary to send individual values for all instances of data. Defaults are sent, then applied to the correct instances, as we discussed previously.
* Lays a foundation for more advanced service management systems enabled by abstraction and encapsulation. These principles allow for the creation of service management applications that can be used by a wide range of people, not just the experts who currently configure most network elements. This reduces operational costs and makes it more cost-effective to develop new services.
* Supports multiple management stations simultaneously. Networks and management stations fail. The SNMP framework has features that permit a managed device to respond correctly to more than one management station at a time. Although this capability has added some complexity to MIB design and management station software, the benefit of supporting access by multiple stations to a single device outweighs the cost. Management stations won't have to maintain continuous TCP connections with each managed device, and can escape

---

*Indeed discussions and meetings continue. There seems to be a consensus that work is needed. There is as yet no consensus about the details of what the improved technology would be.

the inevitable churn that occurs when re-establishing communications following a failure or rollover.

* Provides for hierarchical management and overlapping management areas. Related to the general issue of multiple management systems, SNMPCONF-BACKGROUND document section 4.4 says:

> Networks of size must provide for some level of hierarchical management and overlapping management domains. Often management has a functional flavor where the WAN folks deal with part of the network, and the local people deal with another part of the network. To some extent, even in the policy area there will be overlap. Imagine an edge device which is subject to the policies of multiple administrative domains. Hierarchy is not always exactly straight up and down and it is often the case that policy does not have clean boundaries and systems must to some degree have multiple masters.

* Tightly coordinates objects and their classes. Abstraction is a key element of scale and helps to obscure details from those who don't need them. We must be able to associate instances of configuration objects with their abstract classes for effective service management. When you decide to use SNMP as a management foundation, you get this facility for free. Other approaches have object hierarchies that are not well associated with the instances in managed devices—a real problem if you want to determine whether a configuration change has caused a failure or determine if the increased traffic on specific interface instances is a result of a configuration change or simply an increase in demand.

* Coexists with other management approaches, such as the CLI. An important design goal of the SNMPCONF work was to ensure that the infrastructure supported continuous manageability via other methods as well as SNMP. It was assumed that users would want to maintain the practice of supporting multiple management interfaces on each managed device at the same time. Accordingly, with SNMP-CONF enabled, CLI still functions correctly. That is not necessarily true with other technologies, such as COPS-PR.

Although SNMP has all of these properties, in many respects it was not optimized for every situation.

 # Elements of SNMPCONF

Meeting the requirements listed above required no changes to any portion of the SNMP framework currently in use, but this is not to say that improvements aren't desirable. Work in other areas of the SNMP infrastructure, such as the SMI and protocol efficiency improvements, will certainly enhance the SNMPCONF approach.

SNMPCONF technology consists of two main components: a new MIB Module commonly known as the *Policy-Based Management MIB* [PM], and a script language that is sent to the managed devices in some of the MIB Objects in the PM. These scripts are used to determine which instances in a system are to have a specific policy applied to them and then to apply the defaults appropriate to that policy.

Concepts

The basic concepts of the SNMPCONF approach are simple, but worth making explicit. Let's start with a simple definition:

> Policy-based management is the practice of applying management operations globally on all managed elements that share certain attributes. [PM]

In practice, this means that we want the service management application to send some network devices—selected on the basis of sharing certain attributes—a consistent set of configuration rules for local application to elements (such as interfaces) that have a common set of characteristics.

Policies.   Policies are expressions of the idea that if an element exhibits a certain set of characteristics, we will perform a predetermined operation on that element. Generally, we're talking about setting configuration parameters, but the operation could be anything from terminating a process to collecting statistical information about that element. The PM expresses this relationship as:

```
if (policyCondition) then (policyAction)
```

A *policyCondition* is sent to the managed device in MIB Objects. These MIB objects contain a policy script that is applied to elements like network interfaces to determine if they should be part of a policy. For example the policyCondition might check to see what type each interface in the system is. For those interfaces determined to be of an Ethernet interface type (i.e., the answer is yes to the question posed by the policy-Condition), the *policyAction* is applied to configure the interface in our example to behave in a certain way. Let's step back a minute from this very dense description and see what it means in action.

A PM Module, like other MIB Modules is an SNMP agent* that realizes those MIB Objects belonging to it. As agents go, this module is maybe a little bit smarter than most, and can do more than just read counters or set MIB Object values. It runs 24/7. The PM contains policies, each of which has a condition and an action. The agent is configured with these policy instructions. It also has been configured with a schedule for checking to determine when each policy should be on or off. As we will see, there is more to it than this, but for now, this simple description will do. Operators see only that policies are executed at regularly scheduled times on the devices in their network.

The result of the *policyCondition* is always a boolean value. We can use any combination of MIB Objects and available functions in the policy script language (addition, multiplication, etc.) to define the policy-Condition used to select elements to be operated on by the *policyAction* associated with this specific *policyCondition*. We could consult counter values to look at utilization; the roles assigned to elements (see next section); or the element's basic characteristics, such as *ifSpeed* or *ifType*, and build them into the *policyCondition*, too.

The *policyAction* is an operation that is performed on an element or set of elements selected by the *policyCondition*. An example of the type of work that could be done by a *policyAction* is applying special queuing parameters to those interfaces assigned to important customers so that they get low-latency service.

Elements.   "An element is an instance of a physical or logical entity and is embodied by a group of related MIB variables, such as all the variables for interface #7." [PM] This definition of an element means that with the current SNMPCONF approach, a PM can only manage system parameters and counters that are accessible via SNMP MIB Objects. It is the objective of the localization process we discussed earlier to ensure

---

*There are many different ways to write the software to create SNMP functions in the managed devices. We will discuss this in more detail in the chapters on Designing Management Software for Managed Devices and Specialized Function Subagents.

that those values for any MIB variables that might be set in the policyAction are correct for the specific instance of an object in an element. A good example of this is configuring certain interfaces on a device to give low-latency service to some types of traffic. Because many different MIB objects could potentially control the latency of traffic moving through a device, and these vary from one vendor to another, we want to be sure to send exactly the right set for each interface on each device.*

Roles.   One important characteristic of an element is the role assigned to it by humans using the management system. Some characteristics are not algorithmically determinable by the software, but are important in selecting which parts of a managed device should be part of a specific policy. Roles give the SNMPCONF-enabled system the ability to distinguish between elements of the same type on a system, such as Ethernet interfaces. Roles can be:

* **Political**—An interface that serves the executive offices might be marked with a role "executive," which could be used by a *policyCondition* to select that interface for application of the corresponding *policyAction*. In this case, the Action might be to prioritize packets that arrive on this interface.
* **Financial/legal**—These roles could represent contractual information such as "paid for gold service."
* **Geographic**—Devices west of the Mississippi might be configured with one set of parameters and others in the network with other parameters.
* **Architectural**—Architectural roles are intended to represent network architectural details. They can indicate whether the line is a standby or backup or is a main trunk, or otherwise.
* **Capabilities**—Capabilities constitute another important dimension in our ability to properly localize policy information to the exact devices and elements intended. A capability is represented by an object identifier, most frequently associated with a MIB table. It signals the management system, if present, that the managed device can create or read instances of objects found in the MIB table. The management applications consequently have a better idea of what functions a system is able to perform and can avoid sending policies to systems than cannot implement them. For example, if there are no OIDs in the Capability

---

*The devices should properly deal with requests to change MIB Object values that they do not have. It is far better, however, not to send the incorrect commands in the first place, and the process of localization helps.

Table to indicate that the device supports a particular routing protocol (e.g., BGP), it makes little sense to send policies for BGP.

- **Time**—The SNMPCONF technology recognizes that we may want to take a specific series of policies and download them to a managed device. For a business service that operates one way during peak hours and another during off-peak hours, it can be advantageous to give the managed device both the policies and the schedule that controls their activation and deactivation. A table in the PM provides this scheduling facility. The service management system may then allow the managed device to turn its own policies on and off, but nothing prevents the management system from resuming direct control of the device at any time. It could send new policies to the managed device each time the management system needs to change the device's behavior with regard to a policy. Clearly, this is less efficient than letting the device have all the policies and turn them on and off via local schedules. This feature was included in the design to give flexibility to those who would rather maintain a more direct control of policy changes in their network.

Now that we have covered most of the main concepts, it is easy to see how the distribution of policy to managed devices and subsequent action on that policy works as a simple sequence of actions. The following list is somewhat simplified but is useful as we proceed with a detailed examination of how the SNMPCONF technology works:

- **Step 1**—The management system determines which managed devices to send policies. This may be influenced by its knowledge of the roles and capabilities and other factors known about each device. The management system may also (re)configure roles or restrict capabilities it does not want used on a managed device.*
- **Step 2**—The management system sends the policy information to the managed device. This information includes not only the *policyCondition* and *policyAction*, but also (among other things) information about how often to execute. In this step, the management system sends data that has been localized for the specific device. In the SNMPCONF approach, as we see in step 1, the manager can use information about capabilities and roles to help in this decision-making process. In fact, the management software can use information about

---

*See the role and capabilities tables in the MIB Objects that follow.

the elements known on the device.* Indeed, any retrievable MIB Object can be used to help this process.

* **Step 3**—Once the policy has been installed on the system and turned on, the schedule information takes over. When the schedule is active, the *policyCondition* and *policyAction* scripts execute, according to the information they were configured with.[†]

The Language

When we analyze the requirements for advanced configuration, it turns out that distributing additional intelligence to managed devices is very beneficial. One way of distributing additional intelligence to a device is to program it remotely. In the case of SNMPCONF, this means sending the device instructions in the form of a script. The device uses these instructions, contained in the *policyCondition* and *policyAction*, to select and act on elements in it. In this section, we take a brief look at the SNMPCONF script language and some reasons for its current form. This section is not intended as a primer on the language: for that, the reader is referred to section 6 of the [PM].

Dynamic (Re)Configuration

The use of any language in SNMP will seem unusual to those familiar with the technology. Why not simply design more MIB Objects? A script-based approach has some compelling advantages over adding MIB Objects. One of the most compelling is that scripts are dynamic.

Standard MIB Objects and most representations of configuration information (even XML-based representations) are static. A script, however, can run at intervals to determine which elements in the network device require new configuration information and apply that configuration to them. The values for the configuration of the elements will be on hand and can be applied without operator intervention—a big improvement over traditional configuration techniques.

The actual intervals at which *policyConditions* and *policyActions* run are sent to the device as part of the information in the policy table

---

*See the Element Type Registration Table in the MIB Module that follows.

†See the *pmPolicyConditionMaxLatency* and *pmPolicyActionMaxLatency* objects.

(examined later in this chapter). We could require that *policyCondition* be "evaluated" every 60 seconds for one particular policy and every 2 hours for another. In this context, "evaluated" means that the *policyCondition* script is run, in this case, every 60 seconds. Some events in a device can cause a *policyCondition* evaluation to change, thereby determining whether the corresponding policyAction will be executed. These changes are state changes, utilization changes, and hardware changes.

A *policyCondition* might be written to exempt from the *policyAction* an element found in a disabled or down state that might otherwise have been applied to it. This is an example of an intelligent response to a state change. Putting the smarts into the local device makes dynamic (re)configuration possible. If the disabled element should return to a valid state between executions of a *policyCondition*, it will be reconfigured with the *policyAction* on the next execution of the *policyCondition*.

Sometimes it makes sense to change the configuration parameters for QoS in response to over- or undersubscription or an element. In such a case, we're responding to utilization changes. By including the ability to evaluate information on a periodic basis in the *policyCondition* script, we can effectively change configurations on the fly.

One of the most common problems network operators face is determining the correct configuration for new elements in a network device—hardware changes—and then getting the configuration information installed in a timely fashion. Using the script-based SNMPCONF approach, operators can delegate the job to a policy. The *policyCondition* will check for elements of a given type (Ethernet) and role (connected to the executive suite) during its regular sweeps. When new elements are added to the system, they become visible to the *policyCondition* script and automatically subsumed into the set of elements subject to *policyAction*.* New elements are thus correctly configured by the *policyAction* without ever requiring the operator to investigate policy and log into the machine.[†]

---

*This approach assumes that roles are needed for the policy, and that they were correctly configured when the new element was installed or else roles are not needed for the elements (this may be true a fair percentage of the time) under the control of the policy that manages them.

†Contrast this approach to environments in which traditional customized scripts are used to generate configuration files or CLI configuration commands. The (CLI or File-based) approach is not just time consuming, but error prone, because these types of scripts are modified more frequently than those sent to each device in the SNMPCONF objects. This approach is also less responsive, because these scripts must log in to each device over the network at intervals that will be far less frequent than can be accomplished via a PM, because the PM resides on each device.

Streamlined Data Transmission

Another great advantage of the SNMPCONF scripting approach is the concise transmission of information to, and retrieval of data from the device. Using a script, a device can be handed a set of instructions to determine which elements are to be configured with a particular set of defaults. The device no longer needs the repeated transmission of large amounts of configuration data. This approach also enables the aggregation of statistics based on policies. Although aggregated data collection is not part of SNMPCONF, it follows logically from the initial work.

Take, for example, a policy that selects five elements in a network device—perhaps an interface. We could write a policy that selects these same five elements and then totals the number of packets that have been sent over them for some period of time. The total may be saved in a MIB Object, thus eliminating the need to retrieve potentially many MIB Objects. The efficiency of the entire system is therefore improved.

Familiarity

When creating the SNMPCONF PolicyScript, special care was taken to ensure that the language would be appropriate to the task and at the same time familiar to those most likely to come in contact with it. The [PM] document states:

> Policy conditions and policy actions are expressed with the PolicyScript language. The PolicyScript language is designed to be a small interpreted language that is simple to understand and implement; it is designed to be appropriate for writing small scripts that make up policy conditions and actions.
>
> PolicyScript is intended to be familiar to programmers that know one or more common languages, including Perl and C. PolicyScript is nominally a subset of the C language—however it was desirable to have access to C++'s operator overloading (solely to aid in documenting the language—operator overloading is not a feature of the PolicyScript). Therefore, PolicyScript is defined formally as a subset of the C++ language. A subset was used to provide for easy development of low-cost interpreters of PolicyScript and to take away language constructs that are peculiar to the C/C++ language. For example, it is expected that both C and Perl programmers will understand the constructs allowed in PolicyScript.

Language Characteristics

The language description found in the PM posits features that would be expected in a C/C++/Perl language. PolicyScript has expressions, declarations, statements, variables, operators, and Accessor functions.

These functions are built in for ease of accessing and manipulating SNMP-based information in the managed system. Related functions are grouped together into libraries. A number of policy accessor functions facilitate operation on policies; for example, the *roleMatch()* is used to determine if an element under examination has been assigned a particular role.

Here is an example of what the language looks like, from section 6.3, the PolicyScript QuickStart Guide in the PM:

```
A condition/action pair:
First, register the Host Resources MIB hrSWRunEntry as a new
element in the pmElementTypeRegTable. This will cause the policy
to run for every process on the system. The token '$*' will be
replaced by the script interpreter with a process index (see
Section 7 for a definition of the '$*' token).

The condition:
    // if it's a process and it's an application and it's
    // consumed more than 5 minutes of CPU time
    return (inSubtree(elementName(), "hrSWRunEntry")
            && getVar("hrSWRunType.$*") == 4  // app, not OS or
            driver
            && getVar("hrSWRunPerfCPU.$*") > 30000) // 300
            seconds

The action:
    // Kill it
    setVar("hrSWRunStatus.$*", 4, Integer); // invalid(4) kills it
```

Role of the Language

The PolicyScript language is an essential component of the SNMPCONF technology. Like SNMP, it is not meant to be a human interface and should be hidden from the view of most operators of the service management system. This is consistent with how many networks operate today. A few skilled individuals write and maintain the scripts that interact with CLIs used to configure the managed devices in the network. Each organization that chooses to deploy SNMPCONF technology will develop expertise in this language, however, and over time, service management systems should evolve to generate the majority of PolicyScripts to be sent to managed devices. The fact that the PolicySript is developed to be a standard,* as opposed to a variety of languages designed to interact with a variety of different vendor CLIs, makes possible the development of applications that generate these scripts algorithmically as opposed to the by-hand approach in common use today by most network operators.

The Policy Module and MIB Objects

The second major SNMPCONF technology component is the Policy MIB Module itself. Let's look at it in detail to understand its features and see how they can be used to assist a management system in the configuration and control of services.

The Policy and Policy Code Tables

These tables are the core of the policy information sent to each network device; they contain the essential information for each policy, or at a minimum references to other objects in the PM to control the behavior of the policy.

For those who are not familiar reading MIB Modules, focus your attention on the DESCRIPTION clauses; these contain good information about the functions of each of the objects.

```
-- The policy table

pmPolicyTable OBJECT-TYPE
    SYNTAX       SEQUENCE OF PmPolicyEntry
    MAX-ACCESS   not-accessible
    STATUS       current
    DESCRIPTION
        "The policy table. A policy is a pairing of a
        policyCondition and a policyAction which is used to apply
        the action to a selected set of elements."
    ::= { pmMib 1 }

pmPolicyEntry OBJECT-TYPE
    SYNTAX       PmPolicyEntry
    MAX-ACCESS   not-accessible
    STATUS       current
```

---

*At the time of this writing, PolicyScript is not yet a standard.

```
DESCRIPTION
    "An entry in the policy table representing one policy."
INDEX { pmPolicyAdminGroup, pmPolicyIndex }
::= { pmPolicyTable 1 }


PmPolicyEntry ::= SEQUENCE {
    pmPolicyAdminGroup              UTF8String,
    pmPolicyIndex                   Unsigned32,
    pmPolicyPrecedenceGroup         UTF8String,
    pmPolicyPrecedence              Unsigned32,
    pmPolicySchedule                Unsigned32,
    pmPolicyElementTypeFilter       UTF8String,
    pmPolicyConditionScriptIndex    Unsigned32,
    pmPolicyActionScriptIndex       Unsigned32,
    pmPolicyParameters              OCTET STRING,
    pmPolicyConditionMaxLatency     Unsigned32,
    pmPolicyActionMaxLatency        Unsigned32,
    pmPolicyMaxIterations           Unsigned32,
    pmPolicyDescription             UTF8String,
    pmPolicyMatches                 Gauge32,
    pmPolicyAbnormalTerminations    Gauge32,
    pmPolicyExecutionErrors         Counter32,
    pmPolicyDebugging               INTEGER,
    pmPolicyAdminStatus             INTEGER,
    pmPolicyStorageType             StorageType,
    pmPolicyRowStatus               RowStatus
}


pmPolicyAdminGroup OBJECT-TYPE
    SYNTAX      UTF8String (SIZE(0..8))
    MAX-ACCESS  not-accessible
    STATUS      current
    DESCRIPTION
        "An administratively assigned string that can be used to
        group policies for convenience, readability or to
        simplify configuration of access control.

        The value of this string does not affect policy
        processing in any way. If grouping is not desired or
        necessary, this object may be set to a zero-length
        string."
    ::= { pmPolicyEntry 1 }


pmPolicyIndex OBJECT-TYPE
    SYNTAX      Unsigned32 (1..4294967295)
    MAX-ACCESS  not-accessible
    STATUS      current
    DESCRIPTION
        "An index for this policy entry, unique amongst policies
        in the same administrative group."
    ::= { pmPolicyEntry 2 }


pmPolicyPrecedenceGroup OBJECT-TYPE
    SYNTAX      UTF8String (SIZE (0..32))
```

```
MAX-ACCESS   read-create
STATUS       current
DESCRIPTION
    "An administratively assigned string that is used to
    group policies. For each element, only one policy in the
    same  precedence group may be active on that element. If
    multiple  policies would be active on an element (because
    their  conditions return non-zero), the execution
    environment will only allow the policy with the highest
    value ofpmPolicyPrecedence to be active."
::= { pmPolicyEntry 3 }

pmPolicyPrecedence OBJECT-TYPE
    SYNTAX       Unsigned32 (0..65535)
    MAX-ACCESS   read-create
    STATUS       current
    DESCRIPTION
        "If while checking to see which policy conditions match
        an element, 2 or more ready policies in the same
        precedence group match the same element, the
        pmPolicyPrecedence object provides the rule to arbitrate
        which single policy will be active on this element. Of
        policies in the same precedence group, only the ready and
        matching policy with the highest precedence value (i.e. 2
        is higher than 1) will have its policy action
        periodically executed on this element.

        When a policy is active on an element but the condition
        ceases to match the element, it's action (if currently
        running) will be allowed to complete and then the
        condition-matching ready policy with the next-highest
        precedence will immediately become active (and has its
        action run immediately). If the condition of a higher-
        precedence ready policy suddenly begins matching an
        element, the previously-active policy's action (if
        currently running) will be allowed to complete and then
        the higher precedence policy will immediately become
        active, its action will run immediately and any lower-
        precedence matching policy will not be active anymore.

        In the case where multiple ready policies share the
        highest value, it is an implementation-dependent matter
        as to which single policy action will be chosen.

        Note that if it is necessary to take certain actions
        after a policy is no longer active on an element, these
        actions should be included in a lower-precedence policy
        that is in the same precedence group."
    ::= { pmPolicyEntry 4 }

pmPolicySchedule OBJECT-TYPE
    SYNTAX       Unsigned32 (0..65535)
    MAX-ACCESS   read-create
    STATUS       current
```

DESCRIPTION
        "This policy will be ready if any of the associated
        schedule entries are active.

        If the value of this object is 0, this policy is always
        ready.

        If the value of this object is non-zero but it doesn't
        refer to a schedule group that includes an active
        schedule, then the policy will not be ready, even if
        this is due to a misconfiguration of this object or the
        pmSchedTable."
    ::= { pmPolicyEntry 5 }

pmPolicyElementTypeFilter OBJECT-TYPE
        SYNTAX   UTF8String (SIZE (0..128))
        MAX-ACCESS   read-create
        STATUS   current
        DESCRIPTION
        "This object specifies the element types for which this
        policy can be executed.

        The format of this object will be a sequence of
        pmElementTypeRegOIDPrefix values, encoded in the
        following BNF form:

        elementTypeFilter: oid [';' oid]*
                    oid: subid ['.' subid]*
                    subid: '0' | decimal_constant
        For example, to register for the policy to be run on all
        interface elements, the 'ifEntry' element type will be
        registered as '1.3.6.1.2.1.2.2.1'.

        If a value is registered that does not represent a
        registered pmElementTypeRegOIDPrefix, then that value
        will be ignored."
    ::= { pmPolicyEntry 6 }

pmPolicyConditionScriptIndex OBJECT-TYPE
        SYNTAX   Unsigned32
        MAX-ACCESS   read-only
        STATUS   current
        DESCRIPTION
        "A pointer to the row or rows in the pmPolicyCodeTable
        that contain the condition code for this policy. When a
        policy entry is created, a pmPolicyCodeIndex value
        unused by this policy's adminGroup will be assigned to
        this object.

        A policy condition is one or more PolicyScript
        statements which results in a boolean value that
        represents whether or not an element is a member of a
        set of elements upon which an action is to be performed.

If a policy is ready and the condition returns true for
an element of a proper element type, and no higher-
precedence policy should be active, then the policy is
active on that element.

Condition evaluation stops immediately when any run-time
exception is detected and the policyAction is not
executed.

The policyCondition is evaluated for various elements.
Any element for which the policyCondition returns any
nonzero value will match the condition and will have the
associated policyAction executed on that element unless
a higher-precedence policy in the same precedence group
also matches this element.

If the condition object is empty (contains no code) or
otherwise does not return a value, the element will not
be matched.

When executing this condition, if SNMP requests are made
to the local system, access to objects is under the
security credentials of the requester who most recently
modified the associated pmPolicyAdminStatus object.

These credentials are the input parameters for
isAccessAllowed from the Architecture for Describing
SNMP Management Frameworks[1]."
   ::= { pmPolicyEntry 7 }

```
pmPolicyActionScriptIndex OBJECT-TYPE
     SYNTAX        Unsigned32
     MAX-ACCESS    read-only
     STATUS        current
     DESCRIPTION
```
         "A pointer to the row or rows in the pmPolicyCodeTable
         that contain the action code for this policy. When a
         policy entry is created, a pmPolicyCodeIndex value
         unused by this policy's adminGroup will be assigned to
         this object.

         A PolicyAction is an operation performed on a
         set of elements for which the policy is active.

         Action evaluation stops immediately when any run-time
         exception is detected.

         When executing this condition, if SNMP requests are made
         to the local system, access to objects is under the
         security credentials of the requester who most recently
         modified the associated pmPolicyAdminStatus object.

         These credentials are the input parameters for
         isAccessAllowed from the Architecture for Describing

```
        SNMP   Management Frameworks[1]."
::= { pmPolicyEntry 8 }


pmPolicyParameters OBJECT-TYPE
    SYNTAX       OCTET STRING
    MAX-ACCESS   read-create
    STATUS       current
    DESCRIPTION
        "From time to time, policy scripts may desire one or more
        parameters (e.g., site-specific constants). These
        parameters may be installed with the script in this
        object and are accessible to the script via the
        getParameters() function. If it is necessary for multiple
        parameters to be passed to the script, the script can
        choose whatever encoding/delimiting mechanism is most
        appropriate."
    ::= { pmPolicyEntry 9 }


pmPolicyConditionMaxLatency OBJECT-TYPE
    SYNTAX       Unsigned32 (0..2147483647)
    UNITS        "milliseconds"
    MAX-ACCESS   read-create
    STATUS       current
    DESCRIPTION
        "Every element under the control of this agent is
        re-checked periodically to see if it is under control of
        this policy by re-running the condition for this policy.
        This object lets the manager control the maximum amount
        of time that may pass before an element is re-checked.

        In other words, in any given interval of this duration,
        all elements must be re-checked. Note that it is an
        implementation-dependent matter as to how the policy
        agent schedules the checking of various elements within
        this interval. Implementations may wish to re-run a
        condition more quickly if they note a change to the role
        strings for an element."
    ::= { pmPolicyEntry 10 }


pmPolicyActionMaxLatency OBJECT-TYPE
    SYNTAX       Unsigned32 (0..2147483647)
    UNITS        "milliseconds"
    MAX-ACCESS   read-create
    STATUS       current
    DESCRIPTION
        "Every element that matches this policy's condition and
        is therefore under control of this policy will have this
        policy's action executed periodically to ensure that the
        element remains in the state dictated by the policy.
        This object lets the manager control the maximum amount
        of time that may pass before an element has the action
        run on it.

        In other words, in any given interval of this duration,
```

all elements under control of this policy must have the
action run on them. Note that it is an implementation-
dependent matter as to how the policy agent schedules the
policy action on various elements within this interval."
::= { pmPolicyEntry 11 }

```
pmPolicyMaxIterations OBJECT-TYPE
    SYNTAX        Unsigned32
    MAX-ACCESS    read-create
    STATUS        current
    DESCRIPTION
```
"If a condition or action script iterates in loops too
many times in one invocation, it may be considered by the
execution environment to be in an infinite loop or
otherwise not acting as intended and may be terminated by
the execution environment. The execution environment will
count the cumulative number of times all 'for' or 'while'
loops iterated and will apply a threshold to determine
when to terminate the script. It is an implementation-
dependent manner as to what threshold the execution
environment uses, but the value of this object SHOULD be
the basis for choosing the threshold foreach script.
The value of this object represents a policy-specific
threshold and can be tuned for policies of varying
workloads. If this value is zero, no threshold will be
enforced except for any implementation-dependent maximum.
Regardless of this value, the agent is allowed to
terminate any script invocation that exceeds a local CPU
or memory limitation.

Note that the condition and action invocations are
tracked separately."
::= { pmPolicyEntry 12 }

```
pmPolicyDescription OBJECT-TYPE
    SYNTAX        UTF8String
    MAX-ACCESS    read-create
    STATUS        current
    DESCRIPTION
```
"A description of this rule and its significance,
typically provided by a human."
::= { pmPolicyEntry 13 }

```
pmPolicyMatches OBJECT-TYPE
    SYNTAX        Gauge32
    UNITS         "elements"
    MAX-ACCESS    read-only
    STATUS        current
    DESCRIPTION
```
"The number of elements that, in their most recent
executionof the associated condition, were matched by
the condition."
::= { pmPolicyEntry 14 }

```
pmPolicyAbnormalTerminations OBJECT-TYPE
     SYNTAX        Gauge32
     UNITS         "elements"
     MAX-ACCESS    read-only
     STATUS        current
     DESCRIPTION
          "The number of elements that, in their most recent
          execution of the associated condition or action, have
          experienced a run-time exception and terminated
          abnormally. Note that if a policy was experiencing a
          run-time exception while processing a particular element
          but on a subsequent invocation it runs normally, this
          number can decline."
     ::= { pmPolicyEntry 15 }

pmPolicyExecutionErrors OBJECT-TYPE
     SYNTAX        Counter32
     UNITS         "errors"
     MAX-ACCESS    read-only
     STATUS        current
     DESCRIPTION
          "The total number of times that execution of this
          policy's condition or action has been terminated due to
          run-time exceptions."
     ::= { pmPolicyEntry 16 }

pmPolicyDebugging OBJECT-TYPE
     SYNTAX        INTEGER {
                        off(0),
                        on(1)
                   }
     MAX-ACCESS    read-create
     STATUS        current
     DESCRIPTION
          "The status of debugging for this policy. If this is
          turned on(1), log entries will be created in the
          pmDebuggingTable for each run-time exception that is
          experienced by this policy."
     DEFVAL { off }
     ::= { pmPolicyEntry 17 }

pmPolicyAdminStatus OBJECT-TYPE
     SYNTAX        INTEGER {
                        disabled(0),
                        enabled(1),
                        enabledAutoRemove(2)
                   }
     MAX-ACCESS    read-create
     STATUS        current
     DESCRIPTION
          "The administrative status of this policy.

          The policy will be valid only if the associated
          pmPolicyRowStatus is set to active(1) and this object is
```

set to enabled(1) or enabledAutoRemove(2).

If this object is set to enabledAutoRemove(2), the next
time the associated schedule moves from the active
state to the inactive state, this policy will
immediately be deleted, including any associated entries
in the pmPolicyCodeTable.

The following related objects may not be changed unless
this object is set to disabled(0):
pmPolicyPrecedenceGroup, pmPolicyPrecedence,
pmPolicySchedule, pmPolicyElementTypeFilter,
pmPolicyConditionScriptIndex,
pmPolicyActionScriptIndex,
pmPolicyParameters, and any pmPolicyCodeTable row
referenced by this policy.
In order to change any of these parameters, the policy
must be moved to the disabled(0) state, changed, and
then re-enabled.

When this policy moves to either enabled state from the
disabled state, any cached values of policy condition
must be erased and any Policy or PolicyElement
scratchpad values for this policy should be removed.
Policy execution will begin by testing the policy
condition on all appropriate elements.

[Note to reader: This object exists because a row cannot
sit for extended periods of time with its rowstatus set
to inactive (it is subject to garbage collection). This
object allows policies to be downloaded but not run
except at the convenience of the management station.]"
::= { pmPolicyEntry 18 }

pmPolicyStorageType OBJECT-TYPE
    SYNTAX   StorageType
    MAX-ACCESS   read-create
    STATUS   current
    DESCRIPTION
        "This object defines whether this policy and any
        associated entries in the pmPolicyCodeTable are kept in
        volatile storage and lost upon reboot or if this row is
        backed up by non-volatile or permanent storage."
    ::= { pmPolicyEntry 19 }

pmPolicyRowStatus OBJECT-TYPE
    SYNTAX   RowStatus
    MAX-ACCESS   read-create
    STATUS   current
    DESCRIPTION
        "The row status of this pmPolicyEntry.

        The status may not be set to active if any of the
        related entries in the pmPolicyCode table do not have a

```
          status of active or if any of the objects in this row
          are not set to valid values.

          If this row is deleted, any associated entries in the
          pmPolicyCodeTable will be deleted as well."
    ::= { pmPolicyEntry 20 }

-- Policy Code Table

-- An example of the relationships between the code table and the
-- policy table:
--
-- pmPolicyTable
--      AdminGroup  Index    ConditionScriptIndex
ActionScriptIndex
-- A     ""          1        1                        2
-- B     "oper"      1        1                        2
-- C     "oper"      2        3                        4
--
-- pmPolicyCodeTable
-- AdminGroup   ScriptIndex   Segment     Note
-- ""               1            1        Filter* for policy A
-- ""               2            1        Action for policy A
-- "oper"           1            1        Filter for policy B
-- "oper"           2            1        Action 1/2 for policy B
-- "oper"           2            2        Action 2/2 for policy B
-- "oper"           3            1        Filter for policy C
-- "oper"           4            1        Action for policy C
--

-- In this example there are 3 policies, 1 in the "" adminGroup
-- and 2 in the "oper" adminGroup. Policy A has been assigned
-- script index 1 and 2 (these script indexes are assigned out of
-- a separate pool per adminGroup) with 1 code segment each for
-- the filter and the action. Policy B has been assigned script
-- index 1 and 2 (out of the pool for the "oper" adminGroup).
-- While the filter has 1 segment, the action is longer and is
-- loaded into 2 segments. Finally, Policy C has been assigned
-- script index 3 and 4 with 1 code segment each for the filter
-- and the action.

pmPolicyCodeTable OBJECT-TYPE
    SYNTAX       SEQUENCE OF PmPolicyCodeEntry
    MAX-ACCESS   not-accessible
    STATUS       current
    DESCRIPTION
       "The pmPolicyCodeTable stores the code for policy
        conditions and actions."
    ::= { pmMib 2 }
```

---

*During the evolution of this MIB Module, the term "filter" was replaced by "condition."
So, in the text above think condition where you see phrases "Filter for policy…"

```
pmPolicyCodeEntry OBJECT-TYPE
    SYNTAX       PmPolicyCodeEntry
    MAX-ACCESS   not-accessible
    STATUS       current
    DESCRIPTION
        "An entry in the policy code table representing one code
        segment. Entries that share a common
        AdminGroup/ScriptIndex pair make up a single script.
        Valid values of ScriptIndex are retrieved from
        pmPolicyConditionScriptIndex and
        pmPolicyActionScriptIndex after a pmPolicyEntry is
        created. Segments of code can then be written to this
        table using the learned ScriptIndex values.

        The pmPolicyAdminGroup element of the index represents
        the administrative group of the policy this code entry is
        a part."
    INDEX { pmPolicyAdminGroup, pmPolicyCodeScriptIndex,
            pmPolicyCodeSegment }
    ::= { pmPolicyCodeTable 1 }

PmPolicyCodeEntry ::= SEQUENCE {
    pmPolicyCodeScriptIndex    Unsigned32,
    pmPolicyCodeSegment        Unsigned32,
    pmPolicyCodeText           UTF8String,
    pmPolicyCodeStatus         RowStatus
}

pmPolicyCodeScriptIndex OBJECT-TYPE
    SYNTAX       Unsigned32 (1..4294967295)
    MAX-ACCESS   not-accessible
    STATUS       current
    DESCRIPTION
        "A unique index for each policy condition or action. The
        code for each such condition or action may be composed
        of multiple entries in this table if the code cannot fit
        in one entry. Values of pmPolicyCodeScriptIndex may not
        be used unless they have previously been assigned in the
        pmPolicyConditionScriptIndex or pmPolicyActionScriptIndex
        objects."
    ::= { pmPolicyCodeEntry 1 }

pmPolicyCodeSegment OBJECT-TYPE
    SYNTAX       Unsigned32 (1..4294967295)
    MAX-ACCESS   not-accessible
    STATUS       current
    DESCRIPTION
        "A unique index for each segment of a policy condition
        oraction.

        When a policy condition or action spans multiple entries
        in this table, the code of that policy starts from the
        lowest-numbered segment and continues with increasing
        segment values until ending with the highest-numbered
```

```
        segment."
    ::= { pmPolicyCodeEntry 2 }

pmPolicyCodeText OBJECT-TYPE
    SYNTAX       UTF8String (SIZE (1..1024))
    MAX-ACCESS   read-create
    STATUS       current
    DESCRIPTION
        "A segment of policy code (condition or action). Lengthy
        Policy conditions or actions may be stored in multiple
        segments in this table that share the same value of
        pmPolicyCodeScriptIndex. When multiple segments are
        used, it is recommended that each segment be as large as
        practical.

        Entries in this table are associated with policies by
        values of the pmPolicyConditionScriptIndex and
        pmPolicyActionScriptIndex objects. If the status of the
        related policy is active, then this object may not be
        modified."
    ::= { pmPolicyCodeEntry 3 }

pmPolicyCodeStatus OBJECT-TYPE
    SYNTAX       RowStatus
    MAX-ACCESS   read-create
    STATUS       current
    DESCRIPTION
        "The status of this code entry.

        Entries in this table are associated with policies by
        values of the pmPolicyConditionScriptIndex and
        pmPolicyActionScriptIndex objects. If the status of the
        related policy is active, then this object can not be
        modified (I.E., deleted or set to notInService) nor may
        new entries be created."
    ::= { pmPolicyCodeEntry 4 }
```

Note that objects dealing with administrative groups and precedence in the Policy Table make it possible to group together related policies and to even have one policy take over from another based on predefined conditions.

One of the possible applications for the *pmPolicyAdminGroup* object is to make it easier to identify all the policies that relate to a customer or service. Although the managed system does not use this information in its operation, the values could be used by a management application to give more information to operators. This is very important from an operational perspective because, to deliver a service to customer, it may be necessary to have several policies in place that cover different technologies or operational conditions.

Several possible applications exist for the *pmPolicyPrecedenceGroup* in the service management system. One might group policies for a service together and use the *pmPolicyPrecedence* object to set the priority of each policy in the service. Imagine a case where you have guaranteed a certain bandwidth for traffic from a customer, and one of the interfaces fails. In that case, you might want a backup policy in the pmPolicyPrecedence group that has been designed to produce the desired behavior for this customer in the event of such a failure. This type of dynamic management can only be achieved when we distribute intelligence to the managed devices.

We discuss the schedule table shortly. What is important to notice now about the *pmPolicySchedule* object is that systems can be configured so that the *pmPolicySchedule* table need not be populated. One could then set the *pmPolicySchedule* object to a value of 0 meaning that the policy is always active.

Policies act on elements, and one way to localize policy is to specify the type of element it acts on. Maybe it operates on Fast Ethernet interfaces, or processes that run on a server. The *ServiceElement* class discussed in the chapter on management software* has methods that can add or subtract element types, to send policies to appropriate devices. The *pmPolicyElementTypeFilter* in the Policy Table provides a way for the software executing in the policy module to cause policies to be applied only to those types of elements for which it makes sense.

The *pmPolicyConditionScriptIndex*, *pmPolicyActionScriptIndex*, and Policy Code Table provide a way to send *policyCondition* and *policyAction* scripts to devices that don't fit into an SNMP PDU. Each row in the Policy Code Table can correspond to a part of a single *policyCondition* or *policyAction* script. The two *ScriptIndex* objects are used to associate a specific policy in the Policy Table with entries in the Code Table that contain the actual *policyCondition* and *policyAction* scripts. The management station can send as much information as will fit into a PDU and populate one or more rows in the Policy Code Table. If additional information is required for a single script, the management station can send that information in additional PDUs and create as many rows as needed for the complete *policyCondition* or *policyAction* script. Thus, arbitrarily large *PolicyScripts* can be sent to a managed device, although in practice it is expected that these scripts will be relatively small.

---

*See the figure—Device Relationships in Chapter 11 Management Applications.

Element Type Registration Table

The Element Type Registration Table, in combination with information about the capabilities (see the Capabilities Tables), will be two of the primary methods whereby management applications determine what polices to send to managed devices. Once again, these help the management software in its task of localizing the policy information sent to each managed devices. For example, if we had a device with only Ethernet interfaces (which we could find out from the Element Types Registration Table) and we wanted to get the best performance we could out of the system, we would send one set of configuration parameters. If, on the other hand, there were different types of interfaces in the system, we would send different configuration parameters. In the case where a system had many types of interfaces, we would have to send parameters for each type assuming that at least one of each type was to be configured to deliver the service.

```
-- Element Type Registration Table

-- The Element Type Registration table allows the manager to learn
-- what element types are being managed by the system and to
-- register new types if necessary. An element type is registered
-- by providing the OID of an SNMP object (i.e., without the
-- instance). Each SNMP instance that exists under that object is
-- element. The index of the element is the index part of the
-- a distinct discovered OID. This index will be supplied to
-- policy conditions and actions so that this code can inspect
-- and configure the element.
--
-- For example, this table might contain the following entries,
-- the first three are agent-installed, while the 4th was
-- downloaded by a management station:
--
--   OIDPrefix      MaxLatency Description              StorageType
--   ifEntry        100 mS     interfaces - builtin     readOnly
--   0.0            100 mS     system element - builtin readOnly
--   frCircuitEntry 100 mS     FR Circuits - builtin    readOnly
--   hrSWRunEntry   60 sec     Running Processes        volatile

pmElementTypeRegTable OBJECT-TYPE
    SYNTAX      SEQUENCE OF PmElementTypeRegEntry
    MAX-ACCESS  not-accessible
    STATUS      current
    DESCRIPTION
        "A registration table for element types managed by this
        system.

        Note that agents may automatically configure elements in
```

this table for frequently used element types (interfaces, circuits, etc.). In particular, it may configure elements for whom discovery is optimized in one or both of the following ways:

1. The agent may discover elements by scanning internal data structures as opposed to issuing local SNMP requests. It is possible to recreate the exact semantics described in this table even if local SNMP requests are not issued.

2. The agent may receive asynchronous notification of new elements (for example, 'card inserted') and use that information to instantly create elements rather than through polling. A similar feature might be available for the deletion of elements.

Note that the disposition of agent-installed entries is described by the pmPolicyStorageType object."
```
::= { pmMib 3 }

pmElementTypeRegEntry OBJECT-TYPE
    SYNTAX         PmElementTypeRegEntry
    MAX-ACCESS    not-accessible
    STATUS        current
    DESCRIPTION
        "A registration of an element type."
    INDEX         { pmElementTypeRegOIDPrefix }
    ::= { pmElementTypeRegTable 1 }

PmElementTypeRegEntry ::= SEQUENCE {
    pmElementTypeRegOIDPrefix      OBJECT IDENTIFIER,
    pmElementTypeRegMaxLatency     Unsigned32,
    pmElementTypeRegDescription    UTF8String,
    pmElementTypeRegStorageType    StorageType,
    pmElementTypeRegRowStatus      RowStatus
}

pmElementTypeRegOIDPrefix OBJECT-TYPE
    SYNTAX         OBJECT IDENTIFIER
    MAX-ACCESS    not-accessible
    STATUS        current
    DESCRIPTION
```
"This OBJECT IDENTIFIER value identifies a table in which all elements of this type will be found. Every row in the referenced table will be treated as an element for the period of time that it remains in the table. The agent will then execute policy conditions and actions as appropriate on each of these elements.

This object identifier value is specified down to the 'entry' component (i.e. ifEntry) of the identifier.

The index of each discovered row will be passed to each

invocation of the policy condition and policy action.

The actual mechanism by which instances are discovered is
implementation-dependent. Periodic walks of the table to
discover the rows in the table is one such mechanism.
This mechanism has the advantage that it can be performed
by an agent with no knowledge of the names, syntax or
semantics of the MIB objects in the table. This mechanism
also serves as the reference design. Other
implementation-dependent mechanisms may be implemented
that are more efficient (perhaps because they are
hard-coded) or that don't require polling.
These mechanisms must discover the same elements as the
table-walking reference design.

A special OBJECT IDENTIFIER '0.0' can be written to this
object. '0.0' represents the single instance of the
system itself and provides an execution context for
policies to operate on the 'system element' as well as on
MIB objects modeled as scalars. For example, '0.0' gives
an execution context for policy-based selection of the
operating system code version (likely modeled as a scalar
MIB object). The element type '0.0' always exists - as a
consequence, no actual discovery will take place and the
pmElementTypeRegMaxLatency object will have no effect for
the '0.0' element type. However, if the '0.0' element
type is not registered in the table, policies will not be
executed on the '0.0' element.

When a policy is invoked on behalf of a '0.0' entry in
this table, the element name will be '0.0' and there is
no index of 'this element' (in other words it has zero
length)."
       ::= { pmElementTypeRegEntry 2 }

```
pmElementTypeRegMaxLatency OBJECT-TYPE
      SYNTAX        Unsigned32
      UNITS         "milliseconds"
      MAX-ACCESS    read-create
      STATUS        current
      DESCRIPTION
```
         "The PM agent is responsible for discovering new elements
         of types that are registered. This object lets the
         manager control the maximum amount of time that may pass
         between the time an element is created and when it is
         discovered.

         In other words, in any given interval of this duration,
         all new elements must be discovered. Note that it is an
         implementation-dependent matter as to how the policy
         agent schedules the checking of various elements within
         this interval."
       ::= { pmElementTypeRegEntry 3 }

```
pmElementTypeRegDescription OBJECT-TYPE
    SYNTAX        UTF8String (SIZE (0..32))
    MAX-ACCESS   read-create
    STATUS        current
    DESCRIPTION
        "A descriptive label for this registered type."
    ::= { pmElementTypeRegEntry 4 }

pmElementTypeRegStorageType OBJECT-TYPE
    SYNTAX        StorageType
    MAX-ACCESS   read-create
    STATUS        current
    DESCRIPTION
        "This object defines whether this row is kept
         in volatile storage and lost upon reboot or if this row
         is backed up by non-volatile or permanent storage."
    ::= { pmElementTypeRegEntry 5 }

pmElementTypeRegRowStatus OBJECT-TYPE
    SYNTAX        RowStatus
    MAX-ACCESS   read-create
    STATUS        current
    DESCRIPTION
        "The status of this registration entry."
    ::= { pmElementTypeRegEntry 6 }
```

Role, Capabilities, and Capabilities Override Tables

The Role and Capabilities Tables help a management application localize the most appropriate policies for a particular managed device. As we will see when we get to management software design, information in the Element Type, Role, and Capabilities Tables provide important data used in determining whether a policy is to be sent to a managed device. In combination, these tables incorporate most of what is needed for the management system to make proper policy installation decisions. Where the Capabilities Tables do not provide sufficient granularity with regard to vendor, model, and release restrictions, that information can frequently be gained from other MIB Objects and used to control the installation of policies on the managed device. For instance, some objects might appear in new releases that are not found in previous releases, and these will require configuration. If the management station is aware of the disparity between releases—and it must be, to ensure correct configuration—then it can send the correct parameters to the managed device.

```
-- Role Table

-- The pmRoleTable is a read-create table that organizes role
-- strings sorted by element. This table is used to create and
-- modify role strings and their associations as well as to allow
-- a management station to learn about the existence of roles and
-- their associations.
--
-- It is the responsibility of the agent to keep track of any
-- re-indexing of the underlying SNMP elements and to continue to
-- associate role strings with the element with which they were
-- initially configured.
--
-- Policy MIB agents that have elements in multiple local SNMP
-- contexts need to allow some roles to be assigned to elements
-- particular contexts. This is particularly true when some
-- in  elements have the same names in different contexts and the
-- context is required to disambiguate them. In those situations,
-- a value for the pmRoleContextName may be provided. When a
-- pmRoleContextName value is not provided, the assignment is to
-- the element in the default context.
--
-- Policy MIB agents that discover elements on other systems and
-- execute policies on their behalf need to have access to role
-- information for these remote elements. In such situations, role
-- assignments for other systems can be stored in this table by
-- providing values for the pmRoleContextEngineID parameters.
--
-- For example:
-- Example:
-- element      role    context ctxEngineID #comment
-- ifindex.1    gold                        local, default context
-- ifindex.2    gold                        local, default context
-- repeaterid.1 foo     rptr1               local, rptr1 context
-- repeaterid.1 bar     rptr2               local, rptr2 context
-- ifindex.1    gold    ""      A           different system
-- ifindex.1    gold    ""      B           different system

pmRoleTable OBJECT-TYPE
    SYNTAX       SEQUENCE OF PmRoleEntry
    MAX-ACCESS   not-accessible
    STATUS       current
    DESCRIPTION
        "The role string table.

        The agent must store role string associations in
        nonvolatile storage."
    ::= { pmMib 4 }

pmRoleEntry OBJECT-TYPE
    SYNTAX       PmRoleEntry
    MAX-ACCESS   not-accessible
    STATUS       current
    DESCRIPTION
```

```
                       "A role string entry associates a role string with an
                       individual element."
           INDEX           { pmRoleElement, pmRoleContextName,
                             pmRoleContextEngineID, pmRoleString }
           ::= { pmRoleTable 1 }

PmRoleEntry ::= SEQUENCE {
    pmRoleElement           RowPointer,
    pmRoleContextName       SnmpAdminString,
    pmRoleContextEngineID   OCTET STRING,
    pmRoleString            UTF8String,
    pmRoleStatus            RowStatus
}

pmRoleElement OBJECT-TYPE
    SYNTAX      RowPointer
    MAX-ACCESS  not-accessible
    STATUS      current
    DESCRIPTION
        "The element to which this role string is associated.

        For example, if the element is interface #3, then this
        object will contain the OID for 'ifIndex.3'.

        If the agent assigns new indexes in the MIB table to
        represent the same underlying element (re-indexing), the
        agent will modify this value to contain the new index
        for the underlying element."
    ::= { pmRoleEntry 1 }

pmRoleContextName OBJECT-TYPE
    SYNTAX      SnmpAdminString
    MAX-ACCESS  not-accessible
    STATUS      current
    DESCRIPTION
        "If the associated element is not in the default SNMP
        context for the target system, this object is used to
        identify the context. If the element is in the default
        context, this object is equal to the empty string."
    ::= { pmRoleEntry 2 }

pmRoleContextEngineID OBJECT-TYPE
    SYNTAX      OCTET STRING (SIZE (0..32))
    MAX-ACCESS  not-accessible
    STATUS      current
    DESCRIPTION
        "If the associated element is on a remote system, this
        object is used to identify the remote system. This object
        contains the contextEngineID of the system for which this
        role string assignment is valid. If the element is on the
        local system this object will be the empty string."
    ::= { pmRoleEntry 3 }

pmRoleString OBJECT-TYPE
```

```
    SYNTAX          UTF8String (SIZE (0..64))
    MAX-ACCESS  not-accessible
    STATUS          current
    DESCRIPTION
        "The role string that is associated with an element
        through this table.

        A role string is an administratively specified
        characteristic of a managed element (for example, an
        interface). It is a selector for policy rules, to
        determine the applicability of the rule to a particular
        managed element."
    ::= { pmRoleEntry 4 }

pmRoleStatus OBJECT-TYPE
    SYNTAX          RowStatus
    MAX-ACCESS  read-create
    STATUS          current
    DESCRIPTION
        "The status of this role string."
    ::= { pmRoleEntry 5 }

-- Capabilities table

-- The pmCapabilitiesTable contains a description of
-- the inherent capabilities of the system so that
-- management stations can learn of an agent's capabilities and
-- differentially install policies based on the capabilities.

pmCapabilitiesTable OBJECT-TYPE
    SYNTAX          SEQUENCE OF PmCapabilitiesEntry
    MAX-ACCESS  not-accessible
    STATUS          current
    DESCRIPTION
        "The pmCapabilitiesTable lists the capabilities of a
        system so that policies can be differentially installed
        by management systems based on capabilities.

        Capabilities are expressed at the system level. There
        can be variation in how capabilities are realized from
        one vendor or model to the next. Management systems
        should consider these differences before selecting which
        policy to install in a system."
    ::= { pmMib 5 }

pmCapabilitiesEntry OBJECT-TYPE
    SYNTAX          PmCapabilitiesEntry
    MAX-ACCESS  not-accessible
    STATUS          current
    DESCRIPTION
        "A capabilities entry holds an OID indicating support
        for a particular capability. Capabilities may include
        hardware and software functions as well as the
        implementation of MIBs. The semantics of the OID are
```

defined in the description of pmCapabilitiesType.

Entries appear in this table if any element in the
system has a specific capability. A capability should
appear in this table only once regardless of the number
of elements in the system with that capability. An entry
is removed from this table when the last element in the
system that has the capability is removed. In some
cases, capabilities are dynamic and exist only in
software. This table should have an entry for the
capability even if there are no current instances.
Examples include systems with database or WEB
services. While the system has the ability to create new
databases or WEB services, the entry should exist. In
these cases, the ability to create these services could
come from other processes that are running in the system
even though there are no currently open databases or WEB
servers running.

Capabilities may include the implementation of MIBs but
need not be limited to those that represent MIBs with
one or more configurable objects. It may also be
valuable to include entries for capabilities that do not
include configuration objects since that information, in
combination with other entries in this table, might be
used by the management software to determine whether or
not to install a policy.

Vendor software may also add entries in this table to
express capabilities from their private branch."
    INDEX   { pmCapabilitiesType }
    ::= { pmCapabilitiesTable 1 }

PmCapabilitiesEntry ::= SEQUENCE {
    pmCapabilitiesType   OBJECT IDENTIFIER
}

pmCapabilitiesType OBJECT-TYPE
    SYNTAX   OBJECT IDENTIFIER
    MAX-ACCESS read-only
    STATUS   current
    DESCRIPTION
        "There are three types of OIDs that may be present in
        the pmCapabilitiesType object:

        1) The OID of a MODULE-COMPLIANCE macro that represents
        the highest level of compliance realized by the agent
        for that MIB. For example, an agent that implements the
        OSPF MIB at the highest level of compliance would have f
        the value o '1.3.6.1.2.1.14.15.2' in the
        pmCapabilitiesType object. In the case of software that
        realizes standard MIBs that do not have compliance
        statements, the base OID of the MIB should be used
        instead. If the OSPF MIB had not been created with a

compliance statement, then the correct value of the
pmCapabilitiesType would be '1.3.6.1.2.1.14'. In the
cases where multiple compliance statements in a MIB are
supported by the agent, and one compliance statement
does not by definition include the other, each of the
compliance OIDs would have entries in this table.

MIB Documents can contain more than one MIB. In the case
of OSPF, there is a second MIB in that document that
describes traps for the OSPF Version 2 Protocol. If the
agent also realizes these functions, an entry will also
exist for those capabilities in this table.

2) Vendors should install OIDs in this table that
represent vendor-specific capabilities. These
capabilities can be expressed just as those described.
above for standard MIBs In addition, vendors may install
any OID they desire from their registered branch. The
OIDs may be at any level of granularity, from the root
of their entire branch to an instance of a single OID.
There is no restriction on the number of registrations
they may make, though care should be taken to avoid
unnecessary entries.

3) OIDs that represent one or a collection of
capabilities which could be any collection of MIB
Objects or hardware or software functions may be created
in working groups and registered with IANA. Other
entities (e.g., vendors) may also make registrations.
Software will register these standard capability OIDs as
well as vendor specific OIDs.

If the OID for a known capability is not present in the
table, then it should be assumed that the capability is
not implemented."
    ::= { pmCapabilitiesEntry 1 }

-- Capabilities override table

pmCapabilitiesOverrideTable OBJECT-TYPE
    SYNTAX   SEQUENCE OF PmCapabilitiesOverrideEntry
    MAX-ACCESS not-accessible
    STATUS   current
    DESCRIPTION
        "The pmCapabilitiesOverrideTable allows management
        stations to override pmCapabilitiesTable entries that
        have been registered by the agent. This facility can be
        used to avoid the condition where managers in the
        network send policies to a system that has advertised a
        capability in the pmCapabilitiesTable but which should
        not be installed on this particular system. One case
        that is still in a trial state, or when resources are
        could be newly deployed equipment reserved for some
        other administrative reason. This table can also be used

```
                    to override entries in the pmCapabilitiesTable through
                    the use of the pmCapabilitiesOverrideState object.
                    Capabilities can also be declared available in this
                    table that were not registered in the
                    pmCapabilitiesTable. A management application can make
                    an entry in this table for any valid OID and declare the
                    capability available by setting the
                    pmCapabilitiesOverrideState for that row to valid(1)."
              ::= { pmMib 6 }


pmCapabilitiesOverrideEntry OBJECT-TYPE
      SYNTAX        PmCapabilitiesOverrideEntry
      MAX-ACCESS    not-accessible
      STATUS        current
      DESCRIPTION
              "An entry in this table indicates whether a particular
              capability is valid or invalid."
      INDEX         { pmCapabilitiesOverrideType }
      ::= { pmCapabilitiesOverrideTable 1 }


PmCapabilitiesOverrideEntry ::= SEQUENCE {
      pmCapabilitiesOverrideType              OBJECT IDENTIFIER,
      pmCapabilitiesOverrideState             INTEGER,
      pmCapabilitiesOverrideRowStatus         RowStatus
}


pmCapabilitiesOverrideType OBJECT-TYPE
      SYNTAX        OBJECT IDENTIFIER
      MAX-ACCESS    not-accessible
      STATUS        current
      DESCRIPTION
              "This is the OID of the capability that is declared
              valid or invalid by the pmCapabilitiesOverrideState
              value for this row. Any valid OID as described in the
              pmCapabilitiesTable is permitted in the
              pmCapabilitiesOverrideType object. This means that
              capabilities can be expressed at any level from a
              specific  instance of an object to a table or entire
              module. There are no restrictions on whether these
              objects are from standards track MIB documents or in the
              private branch of the MIB.

              If an entry exists in this table for which there is a
              corresponding entry in the pmCapabilitiesTable, then
              this entry shall have precedence over the entry in the
              pmCapabilitiesTable. All entries in this table must be
              preserved across reboots."
      ::= { pmCapabilitiesOverrideEntry 1 }


pmCapabilitiesOverrideState OBJECT-TYPE
      SYNTAX        INTEGER {
                        invalid(0),
                        valid(1)
                    }
```

```
MAX-ACCESS   read-create
STATUS       current
DESCRIPTION
    "A pmCapabilitiesOverrideState of invalid indicates that
    management software should not send policies to this
    system for the capability identified in the
    pmCapabilitiesOverrideType for this row of the table.
    This behavior is the same whether the capability
    represented by the pmCapabilitiesOverrideType exists
    only in this table, that is it was installed by an
    external management application, or exists in this table
    as well as the pmCapabilitiesTable. This would be the
    case when a manager wanted to disable a capability that
    the native management system found and registered in the
    pmCapabilitiesTable.

    An entry in this table that has a
    pmCapabilitiesOverrideState  of valid should be treated
    as if it appeared in the pmCapabilitiesTable. If the
    entry also exists in the pmCapabilitiesTable in the
    pmCapabilitiesType object, and the value of this object
    is valid, then the system shall operate as if this entry
    did not exist and policy installations and executions
    will continue in a normal fashion."
::= { pmCapabilitiesOverrideEntry 2 }

pmCapabilitiesOverrideRowStatus OBJECT-TYPE
    SYNTAX       RowStatus
    MAX-ACCESS   read-create
    STATUS       current
    DESCRIPTION
        "The row status of this pmCapabilitiesOverrideEntry."
    ::= { pmCapabilitiesOverrideEntry 3 }
```

Schedule Information

The task at hand is to configure the network for secure videoconferencing between Boston and San Francisco. Many network devices in different time zones may require configuration. The problem is that the conference is to take place each Monday, Wednesday and Friday from 3 P.M. EST to 4 P.M. EST. It takes place all year round, even during daylight savings time. For all this configuration activity to transpire correctly, the management system must have a reasonable understanding of each managed element's time zone. Based on that understanding, it must be able to convert the time-related information associated with each policy to be installed on each managed device into hours of the day.

One scalar object, the *pmScheduleLocalTime*, provides a way for the management station to know what local time is for each managed device that supports SNMPCONF technology. The Schedule Table provides a way for the management system to send schedule information to activate and deactivate the policy as needed. Distributing schedule information in this fashion, as opposed to having a management application configure each system every time a policy is to be turned on or off, makes the management task easier, because there is less movement of configuration information between the management station and each of the managed devices. It also could improve the likelihood that videoconferences will take place with the desired QoS and security, since managed devices won't rely on the management application for policy activation and deactivations:

```
-- The Schedule Group

pmSchedLocalTime OBJECT-TYPE
    SYNTAX        DateAndTime (SIZE (11))
    MAX-ACCESS    read-only
    STATUS        current
    DESCRIPTION
        "The local time used by the scheduler. Schedules which
         refer to calendar time will use the local time indicated
         by this object. An implementation MUST return all 11
         bytes of the DateAndTime textual-convention so that a
         manager  may retrieve the offset from GMT time."
    ::= { pmMib 7 }


--
-- The schedule table which controls the scheduler.
--

pmSchedTable OBJECT-TYPE
    SYNTAX        SEQUENCE OF PmSchedEntry
    MAX-ACCESS    not-accessible
    STATUS        current
    DESCRIPTION
        "This table defines schedules for policies."
    ::= { pmMib 8 }

pmSchedEntry OBJECT-TYPE
    SYNTAX        PmSchedEntry
    MAX-ACCESS    not-accessible
    STATUS        current
    DESCRIPTION
        "An entry describing a particular schedule.

         Unless noted otherwise, writable objects of this row can
         be modified independent of the current value of
         pmSchedRowStatus, pmSchedAdminStatus and
         pmSchedOperStatus.  In particular, it is legal to modify
         pmSchedWeekDay, pmSchedMonth, pmSchedDay, pmSchedHour,
```

```
            and pmSchedMinute when pmSchedRowStatus is active."
        INDEX { pmSchedIndex }
        ::= { pmSchedTable 1 }


PmSchedEntry ::= SEQUENCE {
        pmSchedIndex            Unsigned32,
        pmSchedGroupIndex       Unsigned32,
        pmSchedDescr            UTF8String,
        pmSchedTimePeriod       UTF8String,
        pmSchedMonth            BITS,
        pmSchedDay              BITS,
        pmSchedWeekDay          BITS,
        pmSchedTimeOfDay        UTF8String,
        pmSchedLocalOrUtc       INTEGER,
        pmSchedStorageType      StorageType,
        pmSchedRowStatus        RowStatus
}

pmSchedIndex OBJECT-TYPE
        SYNTAX      Unsigned32 (1..4294967295)
        MAX-ACCESS  not-accessible
        STATUS      current
        DESCRIPTION
            "The locally-unique, administratively assigned index for
            this scheduling entry."
        ::= { pmSchedEntry 1 }


pmSchedGroupIndex OBJECT-TYPE
        SYNTAX      Unsigned32 (1..4294967295)
        MAX-ACCESS  read-create
        STATUS      current
        DESCRIPTION
            "The locally-unique, administratively assigned index for
            the schedule group that this scheduling entry belongs to.

            To assign multiple schedule entries to the same group,
            the pmSchedGroupIndex of each entry in the group will be
            set to the same value. This pmSchedGroupIndex value must
            be equal to the pmSchedIndex of one of the entries in the
            group. If the entry is deleted whose pmSchedIndex equals
            the pmSchedGroupIndex for the group, the agent will
            assign a new pmSchedGroupIndex to all remaining members
            of the group.

            If an entry is not a member of a group, its
            pmSchedGroupIndex must be assigned to the value of its
            pmSchedIndex.

            Policies that are controlled by a group of schedule
            entries are active when any schedule in the group is
            active."
        ::= { pmSchedEntry 2 }


pmSchedDescr OBJECT-TYPE
```

```
SYNTAX      UTF8String
MAX-ACCESS  read-create
STATUS      current
DESCRIPTION
    "The human readable description of the purpose of this
    scheduling entry."
DEFVAL { ''H }
::= { pmSchedEntry 3 }
```

```
pmSchedTimePeriod OBJECT-TYPE
SYNTAX      UTF8String (SIZE (0..31))
MAX-ACCESS  read-create
STATUS      current
DESCRIPTION
    "The overall range of calendar dates and times over which
    this schedule is active. It is stored in a slightly
    extended version of the format for a 'period-explicit'
    defined in RFC 2445 [22]. This format is expressed as a
    string representing the starting date and time, in which
    the character 'T' indicates the beginning of the time
    portion, followed by the solidus character '/', followed
    by a similar string representing an end date and time.
    The start of the period MUST be before the
    end of the period. Date-Time values are expressed as
    substrings of the form 'yyyymmddThhmmss'. For example:
```

```
        20000101T080000/20000131T120000
```

```
    January 1, 2000, 0800 through January 31, 2000, noon
```

```
The 'Date with UTC time' format defined in RFC 2445 in
which the Date-Time string ends with the character 'Z' is
not allowed.
```

```
This 'period-explicit' format is also extended to allow
two special cases in which one of the Date-Time strings
is replaced with a special string defined in RFC 2445:
```

```
1. If the first Date-Time value is replaced with the
   string 'THISANDPRIOR', then the value indicates that
   the schedule is active at any time prior to the Date-
   Time that appears after the '/'.
```

```
2. If the second Date-Time is replaced with the string
   'THISANDFUTURE', then the value indicates that the
   schedule is active at any time after the Date-Time
   that appears before the '/'.
```

```
Note that while RFC 2445 defines these two strings, they
are not specified for use in the 'period-explicit'
format. The use of these strings represents an extension
to the 'period-explicit' format."
::= { pmSchedEntry 4 }
```

```
pmSchedMonth OBJECT-TYPE
    SYNTAX       BITS {
                        january(0),
                        february(1),
                        march(2),
                        april(3),
                        may(4),
                        june(5),
                        july(6),
                        august(7),
                        september(8),
                        october(9),
                        november(10),
                        december(11)
                }
    MAX-ACCESS   read-create
    STATUS       current
    DESCRIPTION
        "Within the overall time period specified in the
        pmSchedTimePeriod object, the value of this object
        specifies the specific months within that time period
        that the schedule is active. Setting all bits will cause
        the schedule to act independently of the month."
    DEFVAL { { january, february, march, april, may, june, july,
               august, september, october, november, december } }
    ::= { pmSchedEntry 5 }

pmSchedDay OBJECT-TYPE
    SYNTAX       BITS {
                        d1(0),   d2(1),   d3(2),   d4(3),   d5(4),
                        d6(5),   d7(6),   d8(7),   d9(8),   d10(9),
                        d11(10), d12(11), d13(12), d14(13), d15(14),
                        d16(15), d17(16), d18(17), d19(18), d20(19),
                        d21(20), d22(21), d23(22), d24(23), d25(24),
                        d26(25), d27(26), d28(27), d29(28), d30(29),
                        d31(30),
                        r1(31),  r2(32),  r3(33),  r4(34),  r5(35),
                        r6(36),  r7(37),  r8(38),  r9(39),  r10(40),
                        r11(41), r12(42), r13(43), r14(44), r15(45),
                        r16(46), r17(47), r18(48), r19(49), r20(50),
                        r21(51), r22(52), r23(53), r24(54), r25(55),
                        r26(56), r27(57), r28(58), r29(59), r30(60),
                        r31(61)
                }
    MAX-ACCESS   read-create
    STATUS       current
    DESCRIPTION
        "Within the overall time period specified in the
        pmSchedTimePeriod object, the value of this object
        specifies the specific days of the month within that time
        period that the schedule is active.

        There are two sets of bits one can use to define the day
        within a month:
```

Enumerations starting with the letter 'd' indicate a
day in a month relative to the first day of a month.
The first day of the month can therefore be specified
by setting the bit d1(0) and d31(30) means the last
day of a month with 31 days.

Enumerations starting with the letter 'r' indicate a
day in a month in reverse order, relative to the last
day of a month. The last day in the month can therefore
be specified by setting the bit r1(31), and r31(61) means
the first day of a month with 31 days.

Setting multiple bits will include several days in the
set of possible days for this schedule. Setting all bits
starting with the letter 'd' or all bits starting with
the letter 'r' will cause the schedule to act
independently of the day of the month."

```
    DEFVAL { {   d1, d2, d3, d4, d5, d6, d7, d8, d9, d10,
                 d11, d12, d13, d14, d15, d16, d17, d18, d19, d20,
                 d21, d22, d23, d24, d25, d26, d27, d28, d29, d30,
                 d31, r1, r2, r3, r4, r5, r6, r7, r8, r9, r10,
                 r11, r12, r13, r14, r15, r16, r17, r18, r19, r20,
                 r21, r22, r23, r24, r25, r26, r27, r28, r29, r30,
                 r31 } }
    ::= { pmSchedEntry 6 }


pmSchedWeekDay OBJECT-TYPE
    SYNTAX      BITS {
                    sunday(0),
                    monday(1),
                    tuesday(2),
                    wednesday(3),
                    thursday(4),
                    friday(5),
                    saturday(6)
                }
    MAX-ACCESS  read-create
    STATUS      current
    DESCRIPTION
        "Within the overall time period specified in the
        pmSchedTimePeriod object, the value of this object
        specifies the specific days of the week within that time
        period that the schedule is active. Setting all bits will
        cause the schedule to act independently of the day of the
        week."
    DEFVAL { { sunday, monday, tuesday, wednesday, thursday,
             friday, saturday } }
    ::= { pmSchedEntry 7 }


pmSchedTimeOfDay OBJECT-TYPE
    SYNTAX      UTF8String (SIZE (0..15))
    MAX-ACCESS  read-create
    STATUS      current
```

DESCRIPTION

"Within the overall time period specified in the
pmSchedTimePeriod object, the value of this object
specifies the range of times in a day that the schedule
is active.

This value is stored in a format based on the RFC 2445
format for 'time': The character 'T' followed by a 'time'
string, followed by the solidus character '/', followed
by the character 'T' followed by a second time string.
The first time indicates the beginning of the range,
while the second time indicates the end. Thus, this
value takes the form:
        'Thhmmss/Thhmmss'.

The second substring always identifies a later time than
the first substring. To allow for ranges that span
midnight, however, the value of the second string may be
smaller than the value of the first substring. Thus,
'T080000/T210000'
identifies the range from 0800 until 2100, while
'T210000/T080000' identifies the range from 2100 until
0800 of the following day.

When a range spans midnight, it by definition includes
parts of two successive days. When one of these days is
also selected by either the MonthOfYearMask,
DayOfMonthMask, and/or DayOfWeekMask, but the other day
is not, then the policy is active only during the portion
of the range that falls on the selected day. For
example, if the range extends from 2100 until 0800, and
the day of week mask selects Monday and Tuesday, then the
policy is active during the following three intervals:

        From midnight Sunday until 0800 Monday;
        From 2100 Monday until 0800 Tuesday;
        From 2100 Tuesday until 23:59:59 Tuesday.

    Setting this value to 'T000000/T235959' will cause the
    schedule to act independently of the time of day."
    DEFVAL { '543030303030302F54323335393539'H } --
T000000/T235959
    ::= { pmSchedEntry 8 }

pmSchedLocalOrUtc OBJECT-TYPE
    SYNTAX   INTEGER {
                    localTime(1),
                    utcTime(2)
                 }
    MAX-ACCESS read-create
    STATUS   current
    DESCRIPTION
        "This object indicates whether the times represented in

```
        the TimePeriod object and in the various Mask objects
        represent  local times or UTC times."
    DEFVAL { utcTime }
    ::= { pmSchedEntry 9 }

pmSchedStorageType OBJECT-TYPE
    SYNTAX      StorageType
    MAX-ACCESS  read-create
    STATUS      current
    DESCRIPTION
        "This object defines whether this schedule entry is kept
         in volatile storage and lost upon reboot or if this row
         is backed up by non-volatile or permanent storage.

         Conceptual rows having the value `permanent' must allow
         write access to the columnar objects pmSchedDescr,
         pmSchedWeekDay, pmSchedMonth, pmSchedDay, pmSchedHour,
         and pmSchedMinute."
    DEFVAL { volatile }
    ::= { pmSchedEntry 10 }

pmSchedRowStatus OBJECT-TYPE
    SYNTAX      RowStatus
    MAX-ACCESS  read-create
    STATUS      current
    DESCRIPTION
        "The status of this schedule entry."
::= { pmSchedEntry 11 }
```

Notice that the table establishes the concept of a schedule group, because a single schedule entry might not provide enough flexibility to carry out a business policy. Service management software described later includes a *ScheduleGroup* class for populating this table in a way that ensures consistent operation across a network.

Policies and Elements

Earlier we described some of the data most important for the accurate accounting of services performed on behalf of a customer. These included:

* The total amount of work performed for each customer over specific units of time.*
* A record of outages, or times when the service delivered was less than promised.

---

*In some cases it may be necessary to calculate how much work was performed on a per-instance basis.

Beyond this, accounting for work performed based on services, and the policies that configure the system to realize those services, is critical for effective performance reporting and capacity planning. The SNMP-CONF technology provides tables that make this tracking possible by mapping policies to the specific elements on those devices used to deliver services. By using counter objects on elements that match a specific policy, management systems can obtain critical information they can use to help account for usage by a particular policy. If the policies are constructed so that they are associated with a billable entity, then we have a complete system from configuration to output to a billing system.

```
-- Policy Tracking

-- The "policy to element" (PE) table and the "element to policy"
-- (EP) table track the status of execution contexts grouped by
-- policy and element respectively.

pmTrackingPETable OBJECT-TYPE
    SYNTAX         SEQUENCE OF PmTrackingPEEntry
    MAX-ACCESS  not-accessible
    STATUS         current
    DESCRIPTION
        "The pmTrackingPETable describes what elements
        are active (under control of) a policy. This table is
        indexed in order to optimize retrieval of the entire
        status for a given policy."
    ::= { pmMib 9 }

pmTrackingPEEntry OBJECT-TYPE
    SYNTAX         PmTrackingPEEntry
    MAX-ACCESS  not-accessible
    STATUS         current
    DESCRIPTION
        "An entry in the pmTrackingPETable. The pmPolicyIndex
        in the index specifies the policy tracked by this
        entry."
    INDEX          { pmPolicyIndex, pmTrackingPEElement,
                        pmTrackingPEContextName,
pmTrackingPEContextEngineID }
    ::= { pmTrackingPETable 1 }

PmTrackingPEEntry ::= SEQUENCE {
    pmTrackingPEElement            RowPointer,
    pmTrackingPEContextName        SnmpAdminString,
    pmTrackingPEContextEngineID  OCTET STRING,
    pmTrackingPEInfo               BITS
}

pmTrackingPEElement OBJECT-TYPE
    SYNTAX         RowPointer
```

```
          MAX-ACCESS  not-accessible
          STATUS      current
          DESCRIPTION
              "The element that is acted upon by the associated policy."
          ::= { pmTrackingPEEntry 1 }

pmTrackingPEContextName OBJECT-TYPE
          SYNTAX      SnmpAdminString
          MAX-ACCESS  not-accessible
          STATUS      current
          DESCRIPTION
              "If the associated element is not in the default SNMP
              context for the target system, this object is used to
              identify the context. If the element is in the default
              context, this object is equal to the empty string."
          ::= { pmTrackingPEEntry 2 }

pmTrackingPEContextEngineID OBJECT-TYPE
          SYNTAX      OCTET STRING (SIZE (0..32))
          MAX-ACCESS  not-accessible
          STATUS      current
          DESCRIPTION
              "If the associated element is on a remote system, this
              object is used to identify the remote system. This object
              contains  the contextEngineID of the system on which the
              associated  element resides. If the element is on the
              local system this object will be the empty string."
          ::= { pmTrackingPEEntry 3 }


pmTrackingPEInfo OBJECT-TYPE
          SYNTAX      BITS {
                          actionSkippedDueToPrecedence(0),
                          conditionRunTimeException(1),
                          conditionUserSignal(2),
                          actionRunTimeException(3),
                          actionUserSignal(4)
                      }
          MAX-ACCESS  read-only
          STATUS      current
          DESCRIPTION
              "This object returns information about the previous
              policy script executions.

              If the actionSkippedDueToPrecedence(1) bit is set, the
              last execution of the associated policy condition
              returned non-zero but the action is not active because
              it was trumped by a matching policy condition in the
              same precedence group with a higher precedence value.

              Entries will only exist in this table of one or more
              bits are set. In particular, if an entry does not exist
              for a particular policy/element combination, it can be
              assumed that the policy's condition did not match this
              element."
```

```
::= { pmTrackingPEEntry 4 }
      If the conditionRunTimeException(2) bit is set, the last
      execution of the associated policy condition encountered
      a run-time exception and aborted.

      If the conditionUserSignal(3) bit is set, the last
      execution of the associated policy condition called the
      signalError() function.

      If the actionRunTimeException(4) bit is set, the last
      execution of the associated policy action encountered a
      run-time exception and aborted.

      If the actionUserSignal(5) bit is set, the last
      execution of the associated policy action called the
      signalError() function.



-- Element to Policy Table

pmTrackingEPTable OBJECT-TYPE
    SYNTAX       SEQUENCE OF PmTrackingEPEntry
    MAX-ACCESS   not-accessible
    STATUS       current
    DESCRIPTION
        "The pmTrackingEPTable describes what policies
        are controlling an element. This table is indexed in
        order to optimize retrieval of the status of all
        policies active for a given element."
    ::= { pmMib 10 }

pmTrackingEPEntry OBJECT-TYPE
    SYNTAX       PmTrackingEPEntry
    MAX-ACCESS   not-accessible
    STATUS       current
    DESCRIPTION
        "An entry in the pmTrackingEPTable. Entries exist for
        all element/policy combinations for which the policy's
        condition  matches and only if the schedule for the
        policy is active.

        The pmPolicyIndex in the index specifies the policy
        tracked by this entry."
    INDEX        { pmTrackingEPElement, pmTrackingEPContextName,
                   pmTrackingEPContextEngineID, pmPolicyIndex }
    ::= { pmTrackingEPTable 1 }

PmTrackingEPEntry ::= SEQUENCE {
    pmTrackingEPElement          RowPointer,
    pmTrackingEPContextName      SnmpAdminString,
    pmTrackingEPContextEngineID  OCTET STRING,
    pmTrackingEPStatus           INTEGER
}
```

```
pmTrackingEPElement OBJECT-TYPE
    SYNTAX      RowPointer
    MAX-ACCESS  not-accessible
    STATUS      current
    DESCRIPTION
        "The element acted upon by the associated policy."
    ::= { pmTrackingEPEntry 1 }

pmTrackingEPContextName OBJECT-TYPE
    SYNTAX      SnmpAdminString
    MAX-ACCESS  not-accessible
    STATUS      current
    DESCRIPTION
        "If the associated element is not in the default SNMP
        context for the target system, this object is used to
        identify the context. If the element is in the default
        context, this object is equal to the empty string."
    ::= { pmTrackingEPEntry 2 }

pmTrackingEPContextEngineID OBJECT-TYPE
    SYNTAX      OCTET STRING (SIZE (0..32))
    MAX-ACCESS  not-accessible
    STATUS      current
    DESCRIPTION
        "If the associated element is on a remote system, this
        object is used to identify the remote system. This object
        contains  the contextEngineID of the system on which the
        associated  element resides. If the element is on the
        local system this object will be the empty string."
    ::= { pmTrackingEPEntry 3 }

pmTrackingEPStatus OBJECT-TYPE
    SYNTAX      INTEGER {
                    on(0),
                    forceOff(1)
                }
    MAX-ACCESS  read-write
    STATUS      current
    DESCRIPTION
        "This entry will only exist if the calendar for the
        policy is active and if the associated policyCondition
        returned 1 for this element.

        A policy can be forcibly disabled on a particular
        element  by setting this value to forceOff(1). The agent
        should then act as if the policyCondition failed for
        this element. The forceOff(1) state will persist (even
        across reboots) until this value is set to on(0) by a
        management request. The forceOff(1) state may be set
        even if the entry does not previously exist so that
        future policy invocations can be avoided.

        Unless forcibly disabled, if this value exists its value
```

```
        will be on(0)."
    ::= { pmTrackingEPEntry 4 }
```

Debugging

Policies that support services can fail for many reasons. Having good debugging facilities to trace possible causes for a failure on an element is crucial to maintaining service levels and providing accurate billing information. The Policy Debugging Table assists with this task.

```
-- Policy Debugging Table

-- Policies that have debugging turned on will generate a log
-- entry in the policy debugging table for every runtime exception
-- that occurs in either the condition or action code.

pmDebuggingTable OBJECT-TYPE
    SYNTAX        SEQUENCE OF PmDebuggingEntry
    MAX-ACCESS    not-accessible
    STATUS        current
    DESCRIPTION
        "The pmDebuggingTable logs debugging messages when
        policies experience run-time exceptions in either the
        condition  or action code and the associated
        pmPolicyDebugging object has been turned on.

        It is an implementation-dependent manner as to the
        maximum  number of debugging entries that will be stored
        and the maximum length of time an entry will be kept. If
        entries must be discarded to make room for new entries,
        the oldest entries must be discarded first."
    ::= { pmMib 11 }

pmDebuggingEntry OBJECT-TYPE
    SYNTAX        PmDebuggingEntry
    MAX-ACCESS    not-accessible
    STATUS        current
    DESCRIPTION
        "An entry in the pmDebuggingTable. The pmPolicyIndex in
        the index specifies the policy that encountered the
        exception  that led to this log entry."
    INDEX       { pmPolicyIndex, pmDebuggingElement,
                    pmDebuggingContextName,
pmDebuggingContextEngineID,
                    pmDebuggingLogIndex }
    ::= { pmDebuggingTable 1 }

PmDebuggingEntry ::= SEQUENCE {
    pmDebuggingElement          RowPointer,
    pmDebuggingContextName      SnmpAdminString,
```

```
                    pmDebuggingContextEngineID  OCTET STRING,
                    pmDebuggingLogIndex         Unsigned32,
                    pmDebuggingMessage          UTF8String
    }


    pmDebuggingElement OBJECT-TYPE
        SYNTAX      RowPointer
        MAX-ACCESS  not-accessible
        STATUS      current
        DESCRIPTION
            "The element the policy was executing on when it
            encountered the error that led to this log entry.

            For example, if the element is interface #3, then this
            object  will contain the oid for 'ifIndex.3'."
        ::= { pmDebuggingEntry 1 }

    pmDebuggingContextName OBJECT-TYPE
        SYNTAX      SnmpAdminString
        MAX-ACCESS  not-accessible
        STATUS      current
        DESCRIPTION
            "If the associated element is not in the default SNMP
            context  for the target system, this object is used to
            identify the context. If the element is in the default
            context, this object is equal to the empty string."
        ::= { pmDebuggingEntry 2 }

    pmDebuggingContextEngineID OBJECT-TYPE
        SYNTAX      OCTET STRING (SIZE (0..32))
        MAX-ACCESS  not-accessible
        STATUS      current
        DESCRIPTION
            "If the associated element is on a remote system, this
            object  is used to identify the remote system. This
            object contains the contextEngineID of the system on
            which the associated element resides. If the element is
            on the local system this object will be the empty string."
        ::= { pmDebuggingEntry 3 }

    pmDebuggingLogIndex OBJECT-TYPE
        SYNTAX      Unsigned32
        MAX-ACCESS  not-accessible
        STATUS      current
        DESCRIPTION
            "A unique index for this log entry amongst other log
            entries for this policy/element combination."
        ::= { pmDebuggingEntry 4 }

    pmDebuggingMessage OBJECT-TYPE
        SYNTAX      UTF8String (SIZE (0..128))
        MAX-ACCESS  read-only
        STATUS      current
        DESCRIPTION
```

```
    "An error message generated by the policy execution
    environment. It's recommended that this message include
    the time of day that the message was generated, if known."
::= { pmDebuggingEntry 5 }
```

Notifications

It is essential that the service management system be well synchronized with the state of the devices and services it manages. This allows the service management system to ensure that the desired configuration for each device is properly installed and that the resources of each device are being used as intended.

Devices and the elements they contain and the services that run on them can and do fail. Where there are failures, it is essential that the management system be informed of changes in status. Additionally, new elements which contain new capabilities are constantly added to systems. To assist with these tasks, three notifications have been defined in the policy module:

```
pmNewRoleNotification NOTIFICATION-TYPE
    OBJECTS      { pmRoleStatus }
    STATUS       current
    DESCRIPTION
        "The pmNewRoleNotification is sent when an agent is
        configured with its first instance of a previously unused
        role string (not every time a new element is given a
        particular role).

        An instance of the pmRoleStatus object is sent containing
        the new roleString in its index. In the event that two or
        more elements are given the same role simultaneously, it
        is an implementation-dependent matter as to which
        pmRoleTable  instance will be included in the
        notification."
    ::= { pmNotifications 1 }

pmNewCapabilityNotification NOTIFICATION-TYPE
    OBJECTS      { pmCapabilitiesType }
    STATUS       current
    DESCRIPTION
        "The pmNewCapabilityNotification is sent when an agent
        gains a new capability that did not previously exist in
        any  element on the system (not every time an element
        gains aparticular capability).

        An instance of the pmCapabilitiesType object is sent
        containing  the identity of the new capability. In the
        event that two or more elements gain the same capability
        simultaneously, it is an implementation-dependent matter
```

```
                        as to which pmCapabilitiesType instance will be included
                        in the notification."
            ::= { pmNotifications 2 }

        pmAbnormalTermNotification NOTIFICATION-TYPE
            OBJECTS      { pmTrackingPEInfo }
            STATUS       current
            DESCRIPTION
                "The pmAbnormalTermNotification is sent when a policy's
                pmPolicyAbnormalTerminations gauge changes value from
                zero to any value greater than zero and no such
                notification has been sent for that policy in the last 5
                minutes.

                The notification contains an instance of the
                pmTrackingPEInfo  object where the pmPolicyIndex
                component of the index identifies the associated policy
                and the rest of the index identifies an element on which
                the policy failed."
        ::= { pmNotifications 3 }
```

In some implementations, additional notifications could be created when there are configuration changes to the PM or other objects in the system. Circumstances may arise when it is necessary for one or more devices to be configured by hand, using a CLI or other access method. In these cases, it is essential that the service management system be informed about changes to the configuration of a device by the CLI. If this is not done, we no longer know how each device is really configured and, among other issues, we lose whatever ability we had to produce accurate bills for those elements that have been reconfigured without the management systems' knowledge. We must accomplish our goal of keeping the management system in sync with changes made by the CLI without placing additional requirements on the users of the CLI or other access methods. Although not part of the base SNMPCONF technology, a well designed system (whether it uses SNMPCONF or not) should send notifications to its designated management systems whenever changes in configuration are made, regardless of the access method.* These notifications must be designed not to overwhelm the management system with gratuitous information. (We discuss additional issues about supporting multiple access methods in the chapter on Infrastructure Design Considerations.)

---

*SNMPCONF does provide a mechanism to exempt an element that would otherwise be covered in a policy. This would allow CLI or other mechanisms to make changes to this element and avoid the possibility that CLI will cause a conflict with a policy or a policy overriding an entry put in by an operator via the CLI. See the *pmTrackingEPStatus* object in the Element to Policy Table.

A Complete System

When fully implemented, along with the configuration objects that are operated on by the PM module, SNMPCONF technology can be used effectively by a service management system to ensure reliable high-value services.

This system has a coordinated view, in a single name space, of all the management objects within it. This includes all fault, configuration, accounting, and performance data. Even when SNMPCONF is used, however, management software retains access to all the individual MIB Objects in a system (assuming it has the correct access privileges). The manager can retrieve fault, security, performance, and other accounting data directly from the system, even if it is not under the SNMPCONF PM umbrella. The management system will know from which elements to collect the data, as a result of the information it can retrieve from the Policy Tracking and Element to Policy Tables. This is essential, because we want to make sure that we have a complete hierarchy in the system; that is, we need to map policies from the highest level of abstraction all the way down to specific elements. The management software will also have the ability to abstract and encapsulate information, which can improve the efficiency of the system and make it possible for more people to use the management system. This complete hierarchy of abstraction and encapsulation is useful because of the ability to move up and down the hierarchy, all the way down to the element level. This feature is something none of the previously described approaches to policy-based management can claim.

PART

4

Designing
Management
Software

Our focus in Chapter 8 is on the design choices available to engineers as they create management software for modern network systems. Particular attention is paid to the issues that arise when multiple access methods such as SNMP and CLI are supported on the same device. Chapter 9 considers some existing IETF MIB Modules that can be used with the system described in Chapters 7 and 8, and demonstrates how configuration and other management information can be connected in an SNMP based service management system.

CHAPTER 8

Designing Management Software for Managed Devices

Effective management is not free. Costs must be borne equally by the managed devices and the management applications. For too long, too many of us pushed the problem of management to external Perl scripts, management applications, and chance. The level of management software deployed in managed devices is particularly poor, especially considering that these devices play a key role in an effective service management system; in fact, they are its foundation. Without good management control and instrumentation in the managed devices, there is no way to achieve a good service management system. This is true regardless of how much time and energy is devoted to management applications.

You've already seen the benefits of a service management system based on a single name space, which implies a single management framework. Having identified the factors important to effective service management, we concluded that SNMP is a reasonable foundation on which to build a service management infrastructure. In reality, you can expect to see ongoing customer demand for many different frameworks, including SNMP, CLI, and configuration file-based access, Web, and more. Hence, this chapter and those that follow are presented with the goal of developing a system that works optimally when managed devices support multiple access methods. This same philosophy is carried through to the chapters that deal with service management applications.

Service Management Components Inside Managed Devices

Figure 8.1 identifies the major components of the portions of a service management system that resided on a managed device.

The implementation details and relationships between these various components either fosters or impairs effective management.

The access layer represents all possible ways that the managed element might be accessed for management purposes. This layer can be divided into parts, each representing one of the access protocols and related security infrastructure. SNMP typically provides its own security and other controls. In the case of SNMP, these controls are an integral part of the framework, whereas in the case of the CLI they are external and are provided by mechanisms such as SSH® [SSH].

The middle layer of the diagram, access methods to operational instrumentation, represents all the code written by either the SNMP agent or CLI developers to access and process information at the layer below. In a

Figure 8.1
Managed device
service management
software
components.

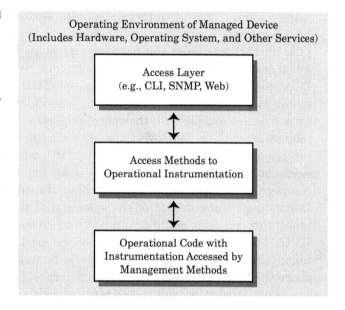

Operating Environment of Managed Device
(Includes Hardware, Operating System, and Other Services)

Access Layer
(e.g., CLI, SNMP, Web)

Access Methods to
Operational Instrumentation

Operational Code with
Instrumentation Accessed by
Management Methods

device that implements the System Application MIB Module [RFC2287], for example, the agent software that implements this MIB Module contains code developed for the specific environment in which the agent executes. The software is designed to query the operating system about the memory and CPU utilization of different software running on the system. When requests arrive in the managed device for MIB Objects on CPU or memory utilization (objects found in the System Application MIB Module), this code can make the appropriate query and return the correct result.

The lowest layer in the diagram is potentially the most important and yet generally outside the control of those developing the management software. This layer represents the operation code in a device, and the user-visible and valued functions. If the operational software and the hardware on which it executes have not been designed with management in mind, many difficulties ensue. The issues, best addressed by the software and the underlying hardware of the system as it is being designed, include:

- **Memory**—The amount of memory available for the retention of counter and other information by the operational software. The operational software must hold this information until the middle layer retrieves it, thereby freeing up space for the operational code to store more information. This simple yet critical factor has a significant impact on the management software. In a system with high-speed interfaces, in which the operational code is often supported by hardware counters, these counters have finite available memory. If insufficient memory is provided by the system, then operational code must pick up the data from the hardware at more frequent intervals to avoid losing information. The management software at the middle layer must also pick up data from the operational code at a more frequent rate, depending, of course, on how much memory and other resources are available for the operational code to devote to storage of management data. Moreover, operational code often executes on the interface card, as opposed to the main CPU, where much of the management software will execute. This is an important issue since the management data must move across the internal system bus. Depending on the design of the system and the amount of data to move, this can represent a significant load on the bus.

- **CPU**—We know it's undesirable to place most of the complexity of management on applications, as opposed to making managed devices share the load. We've seen the negative impact of a design that places too much burden on one component, and this same problem exists inside the managed device. As systems get larger and more complex, it makes more sense to have them perform more work. We'd like them to take over some statistics summarization, verification of configuration data, and improve the quality of asynchronous notifications (e.g., notification messages) from a managed device. All of these techniques require CPU and memory. Some techniques result in higher CPU utilization by the operational as well as management code.

- **APIs/access**—Management software cannot control what it cannot access. If the underlying code does not provide a good interface for management software, the valuable data in the operational code will not be accessible. Similarly, if the operational code has an inefficient interface, it may still be able to access the data, but only at too high a cost to overall system performance. A smart question to ask vendors when evaluating their products is how many management operations of a particular type their system can support before services on that system are adversely affected. If the number is below your requirements, either ask for an improvement or look for another vendor.

- **Granularity of control**—Closely related to the APIs and access facilities of operational code is the level of granularity supported by the code. Granularity is how much of a system or subsystem can be reconfigured at one time with a single operation. Later in this chapter we discuss additional details about the granularity of configuration control. For now, we need to know the API exposes a level of granularity that allows management software access at a meaningful level.

When configuration is expensive, we often look first to an inefficiency in the operation code, but it may also be the result of very coarse configuration changes that require a great deal of reinitialization before services can be restored. The most extreme case of this is requiring a system to reboot after a change.

* **Transactions and reconfiguration speed**—Transaction support is also related to granularity of control. Does your management software support very fine- to coarse-grained transactions? It makes little difference if the operational software can only deal with very coarse transactions—such as rereading an entire configuration file. (The exception to this rule is when you are trying to reduce service interruptions, in which case you many want to store up many configuration changes in the management software that executes on managed devices and then pass these changes on to the operational software in one large chunk.)

For readers who are not involved in developing operational or management code that executes inside managed devices, these issues may seem somewhat removed from day-to-day concerns, but that impression is far from true. A system's ability to update counters affects how often we can meaningfully poll for data and get useful information. Many management systems offer very fine-grained polling, even down to the level of once per second. The problem is that if the managed device can't update counter information as frequently, the polls won't provide any additional information. They will, however, represent needless load on the management application, network, and managed devices.

An analogous problem exists for the folks who configure managed devices. With poorly designed management software on the devices, or poorly designed operational code underlying them, configuration changes can be quite expensive. This is true regardless of the method(s) used to convey the configuration change to the managed device. In such systems, even a small change may require restarting an entire subsystem, which will result in a service interruption. If that's your situation, then your configuration approach will be different from the one you'd take if changes could be easily supported by the operational code and underlying hardware.*

---

*The nature of the work that the operational code performs must be taken into consideration since some configuration parameters are better changed as an integrated set. Failure to do so can result in improper behavior of the service. Future standards should place greater emphasis on the specification of configuration parameters, those that can/should be changed together, and what the operational behavior of the system should be during the reconfiguration operation. See the discussion on transaction integrity in the operational software later in this chapter.

Infrastructure Design Considerations

In the previous section, we focused primarily on how management software can be integrated into a managed device and how the operational software in that device can affect manageability. In this section, we discuss specific management software design considerations that address those issues. Well-designed software will:

- Support multiple access methods and the convergence layer
- Deal with large and distributed devices
- Ensure transactional integrity
- Accommodate counters
- Effectively deal with notifications

Supporting Multiple Access Methods and the Convergence Layer

Designers of systems that must support multiple access methods have an immediate choice and then a second choice that flows from it. The first choice is whether to integrate the disparate access methods and create a convergence layer where a common name space is used. Assuming you opt for an integrated environment, you must decide where to place the convergence layer.

Figure 8.2 illustrates a nonintegrated ships-in-the-night approach taken by some vendors, in which each access method has its own access methods to instrumentation. The figure shows only SNMP and CLI, but you could include as many parallel paths into the device as you want, subject to the limitations of physical resources, performance, and development and testing resources.

In the diagrams to follow, blocks that contain "SNMP Access" or "CLI Access" are intended to represent the part of those respective systems that interacts with the world outside the managed device. For SNMP, this would be the part of the system that receives an SNMP request or sends a response or notification. The "brains" in Figure 8.2 are in the boxes marked "access methods." For SNMP, this code receives the request from the SNMP access layer and performs the real work of retrieving data and processing it.* The CLI side of the diagram behaves similarly: the CLI access layer interacts with the outside universe and may parse the request. The code that interacts with the device's operational code is found in the box directly below the CLI Access box.

In Figures 8.2 and 8.3, the layers are nice and clean. That's often not true of real systems, especially in those environments that are older and were developed when architectural issues seemed simpler and when multiple access layers were less of a concern. The layered approach becomes more critical when a convergence layer is introduced.

Figure 8.2
Independent access layers.

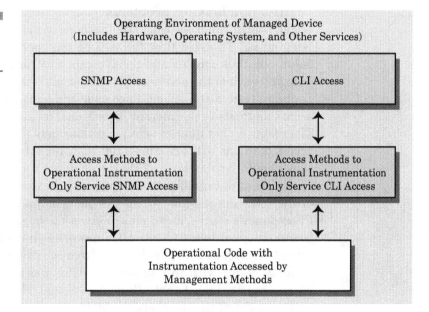

This ships-in-the-night approach has some disadvantages, but also some significant benefits, including faster implementation of the access layers. Because each access layer is independent, you won't spend time coordinating with some mediation layer, and converting name spaces, and juggling all the issues that arise from different name spaces. Because these are independent implementations, by definition, the vendor can ship one access method before the other, based on customer demand. And because these are simpler implementations, you can sometimes achieve better performance than you can by tying together multiple access methods.

---

*For a more detailed view of the general SNMP architecture see "An Architecture for Describing SNMP Management Frameworks" [RFC2571].

But if there are such strong benefits to this approach, there are also strong negatives :

- **Duplicate development and testing of the multiple access methods**—It is obvious that each stack must be implemented. What is not always obvious is that the ships-in-the-night approach also requires us to implement for each stack the code that accesses operational code. Each of these separate code bases must also be tested.
- **Transaction/locking problems**—During the time a configuration activity is in progress with one method, it's possible for another to "step on it." To avoid this, the API to the operational code must be enhanced, thus causing additional complexity for both the operational code and each of the access methods that must use the locking API.
- **Inconsistent data**—Sometimes code implemented in two different access methods uses different algorithms, or just has different bugs. The result is that queries sent simultaneously to a device by different access methods will sometimes return values that are different when they should be the same. This result doesn't automatically indicate inconsistent data; different answers may have to do with the timing of the code and its state when the query is made. Regardless of the discrepancy's source, it can make network debugging more difficult.
- **Security problems**—No two frameworks share the same security infrastructure, so it is possible to introduce inadvertent holes in your security policy with this model. SNMPv3 has a sophisticated user-based security model that allows access control based on user group and specific access to specific objects in the system. Name space issues make mapping hard, so it is possible to have security errors (access where an individual is not meant to have access). A more direct and significant threat is basic misconfiguration. The fact that each access system must be configured is yet another operational cost of this approach.
- **Extra hardware resources**—In the end, two complete stacks usually take more of everything than one.

Now suppose you've made the decision to converge the stacks. Where is the convergence layer to be placed? There are generally two options, each having its pros and cons. The first is to create a common layer just above the operational code. The common layer could, in a sense, be an extension of the API for the operational code and might look something like Figure 8.3.

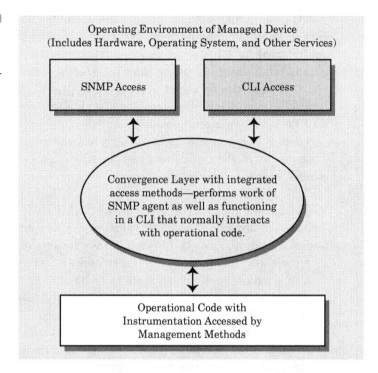

When an SNMP agent implements a MIB object, the code that it writes calls this convergence layer and, in turn, the convergence layer interacts with the operational code to obtain and process information. The information is then returned to the agent, which only has to package up the information for a response. It is possible to have the processing of the information done at the SNMP or CLI layers, but this diminishes the value in having the convergence layer, because you will have to write the processing code for each. It is better to have the convergence layer do as much processing as possible with this design approach to avoid duplicate coding—one of the reasons for creating the convergence layer in the first place.

If the convergence layer also supports asynchronous notifications, it can meet the requirements for an SNMP access layer, which must, from time to time, send out asynchronous messages. The SNMP access layer gets the information it needs for the generation of these notifications from the convergence layer, which can be designed in a number of ways to send asynchronous messages.

The convergence layer performs whatever monitoring actions are necessary to send a message to the SNMP agent. The SNMP agent then sends a notification depending on the MIB Object definitions supported by the agent. For the CLI, the behavior is similar, except we do not normally think of CLIs as supporting asynchronous events. When a CLI command is entered, rather than interacting with the operational code, the CLI calls the same interface used by the SNMP agent; the convergence layer performs the necessary work and returns the results to the CLI for formatting and presentation.

Placing the common layer directly above the operational code eliminates many of the difficulties associated with maintaining unintegrated access methods. Specifically:

- The access methods in the CLI or SNMP code are simplified. They access operational instrumentation, as described in Figure 8.2, as Independent Access Layers. These access method functions are moved into the convergence layer and are accessible to the now simplified access methods in the CLI or SNMP (or whatever a protocol is used to access the device). Here, we save and implement the code that accesses the instrumentation only once.
- Some of the overhead for two different stacks is reduced, although there is some overhead for each stack to translate into and out of the canonical form represented by the convergence layer.
- Some of the testing overhead is reduced.

Another choice for the convergence layer, and one preferable in many respects, is to select one access method and use its name space and access methods as the convergence layer. Commercial tools for this sort of design are available from several vendors. Systems using SNMP as the convergence layer have been successful with this approach; such a system is shown in Figure 8.4.

When SNMP acts as a convergence layer for other access methods like a CLI, the SNMP side behaves as if it were operating in a ships-in-the-night environment. An SNMP agent implements all the code necessary to perform all the functions usually assigned to an SNMP system. This includes all the code for the "access methods to instrumentation" that was moved to the convergence layer in Figure 8.3. Here, the convergence layer is placed on top of the SNMP agent, which allows non-SNMP systems to make full use of the SNMP infrastructure in the managed element.

Because this layer is below the protocol access layer, there is no SNMP protocol involved. The other access layers convert to the SNMP name space. For example, a CLI command that displays counters would be parsed in the normal way by the CLI code. But once the CLI system knows what the command means, it will be converted to the SNMP name space (OIDs) so that it can be sent via an API to the convergence layer. The same methods used by an SNMP agent to support an SNMP request are used to perform the desired operation on the desired objects. Upon completion of the operation, the SNMP system returns the results through the convergence layer back to the CLI, where the results are formatted and presented.*

Figure 8.4
Convergence at the
SNMP access layer.

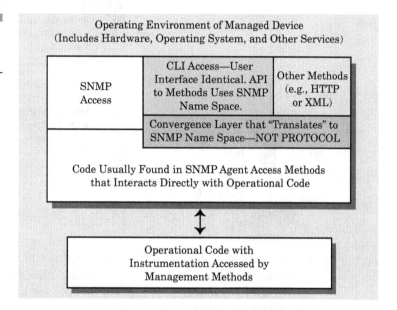

Figure 8.5 (commonly called a *sequence diagram*) shows a sequence of events from a CLI request to the return of results from the convergence layer for processing by the CLI. This sequence shows a possible interaction when using the architecture shown in Figure 8.4. Although our example is a CLI, other access methods can work as well. The benefits of this approach include all those benefits associated with a convergence layer above the operational code plus:

▪ **Transaction control**—Convergence at this layer makes it is easier for multiple access methods to coexist without stepping on each other

---

*Note that in the types of systems described here, there are no SNMP PDUs created inside the managed device when making these name space translations, only conversion to the SNMP name space—that is, conversion to the SNMP object or objects.

Figure 8.5
CLI convergence
with the SNMP
access layer.

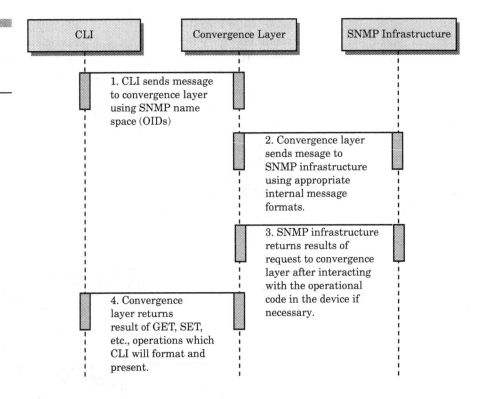

at a transaction level. If a CLI command needs to access or change a value, the SNMP infrastructure locks that resource to avoid collisions, until the transaction is completed or fails for some reason. In cases where the CLI needs many resources, which is not an unusual circumstance, the CLI sends the message with a request for all the resources; those available are then locked until this transaction completes.

If the convergence layer is placed elsewhere, it must implement a transaction control system.

Implementation is likely to be complex and time consuming. Furthermore, it requires the system to understand the transaction semantics of each access method and mediate requests, or at least to deny resources consistently and gracefully.

■ **Security infrastructure coordination**—Because we're assuming an SNMP infrastructure, we know a consistent security approach can be implemented, although this need not be visible to users. It can be used to help ensure that access to a device's management controls is consistently configured, based on security policy. Figure 8.6 shows how it would look.

Figure 8.6
Security convergence at the SNMP layer.

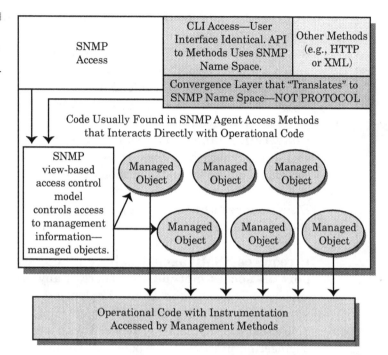

Each of the Managed Objects in the diagram represents a piece of management information accessible via SNMP. These Managed Objects access information in the operation code according to their definition and implementation. The SNMP framework provides tools by which users can be assigned to groups with certain access privileges associated with specific object. Mapping CLI access to data elements in the SNMP framework is relatively straightforward and has been accomplished in a number of commercial implementations. When a user logs onto the CLI with a password, he has the same access to data as when using SNMP—assuming of course that the user has the correct access privileges. Of course, if you want to introduce differences in access privileges based on the access methods, this can be accommodated.

* **Assistance from third-party code**—Some third-party SNMP infrastructures come with APIs in place that use this approach to support access methods in addition to SNMP. A common access supported by this third-party code is Web-based access. Tools are provided to help developers (and sometimes users) map visible Web pages to SNMP objects that are passed through the convergence layer and the security system in the same way as the CLI example in the previous

sequence diagram. In this case, developers not only save the cost of writing the convergence layer, they save the cost of writing an API to the convergence layer. True, they may sometimes want to adapt it, but this is often easier than starting from scratch. Because the CLI is so important to most network equipment, a great deal of attention is being paid to further improving the services of the API to the convergence layer offered to CLIs.

▨ **Converging name space sooner rather than later**—This allows each access method to convert in a contextually correct way to the common name space and operational semantics.

Disadvantages associated with using an SNMP-based infrastructure as the convergence layer are:

▨ **It requires SNMP agents to have all objects necessary to support all access methods**—In short, if any MIB Objects needed to support a CLI command are not implemented, the CLI command will not work. The time lag from MIB Object implementation to CLI availability can be significantly reduced with a good infrastructure and engineering practices. The potential for such a time lag is a significant concern in most environments, and the problem is not unique to SNMP or CLI. Any infrastructure used to support all the other access methods has the same requirement. One can migrate to this design over time, leaving the other ships-in-the-night stacks in place—or at least portions of them.

▨ **It may tie implementation to a specific supplier of an SNMP infrastructure**—The convergence layer will necessarily have to be tied to some SNMP infrastructure implementation, because there is no open standard available for CLIs or other access approaches using the internals of an SNMP agent. If you choose to move to another vendor, you may encounter significant recoding to integrate your convergence layer with the new infrastructure. How hard that turns out to be will be a function of the design of the infrastructures and the convergence code.

▨ **Only those operational semantics supported by SNMP can be supported**—In some cases, extra code will be needed at the convergence layer or in the CLI for the conversion of semantics in an access layer that is not supported by SNMP. For example, some CLI systems have a Commit command that can span many operations. Although SNMP does have some primitives to help, these are limited at the SNMP operations level (e.g., GET, SET, etc.) From an implementation perspective, additional objects may have to be defined for these circumstances to help with large-scale commits. It is also possible to collect a series of commands at the conversion layer and wait until the explicit CLI commit is made, then send all the requests at once to the SNMP layer for processing.

None of these alternatives is difficult, and it's often a part of building a system with an effective approach to transaction control. Also, note that when converging using a separate layer just above the operational code, the code must be modified to support and convert all the semantics necessary to all the different access methods. Failing this, each access method must provide the conversion in its interface to the convergence layer. In either case, a cost is added in supporting multiple different access methods—a necessary but suboptimal situation if you want to support multiple user access methods for a system.

In the end, so many factors can influence your decision about whether to integrate multiple access methods that I can't offer much specific advice. In cases where there is an extant system in place, the best technique may be a hybrid approach that allows migration from a less to a more integrated system. Some of the variables that must be considered include:

* Time to market
* Human resources available and the skills they possess
* Access methods appropriate to the product
* Customer influence; many customers often express strong opinions about design issues for the products they consider for purchase
* Performance considerations and hardware resources available

Dealing with Large Distributed Products

We've challenged our vendors to achieve high levels of performance for the services their products offer. These services can be anything from packet forwarding to reliable Web services to bandwidth management and QoS. Functions are sometimes combined in a single device, which adds a significant level of complexity to the problem. To address this challenge, vendors may want to use multiple CPUs to distribute the load. The problem with this approach is that, to save hardware expense and software complexity, these CPUs typically operate as independent systems with little knowledge of what each is doing. Sure, they exchange messages, but often in a very loosely coupled fashion. Look inside many of these large, complex systems and they closely resemble heterogeneous local area computer networks.

Figure 8.7 shows a common generic system. Each of the boxes in the shaded area represents a separate computer system. Depending on the architecture, each box could be performing a separate function, backing up a different system inside the box, or performing load-sharing tasks with another system in the box. The dashed lines represent the *back-plane* of the system; in other words, the hardware and software infra-structure by which one system communicates with another inside the device. These back-planes often use Ethernet technology because of its flexibility, range of speeds, and widely available building block compo-nents (e.g., hubs and switches). Figure 8.7 shows two internal interfaces on each system, connected with one another via a hub/switch. This approach is commonly used to provide redundancy.

Figure 8.7
A complex system.

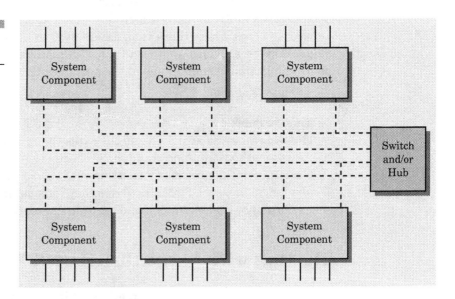

——— External Interfaces

· - - - Internal Interfaces Used for System Component Communication

Will it work? The problem with all this distribution, complexity, and heterogeneity is that end users (for the most part) do not want to see it. They are more interested in the intended benefits such as high reliabili-ty, throughput, and advanced functions. Complexity creates problems for whatever management software is to be distributed inside the computing environment. Design choices made in the previous section affect choices to be made in this section. They in turn determine how easy it is to manage devices of this kind, and how cost-effective it will be to deploy value-added services on them. Some user-related issues are:

* **A single (primary) address for management**—Figure 8.7 shows multiple external interfaces on each of the systems. (An external interface is designed for the system to communicate with the rest of the network, as opposed to one dedicated to communication between individual internal components.) Not all system components in this large distributed box will have external facing interfaces, or the same number of them. Nevertheless, it is common for many system components to have several interfaces that can communicate with the rest of the network. The problem with multiple interfaces, from a management perspective, is that users would like a single IP address—the management address (perhaps with a backup or two)—to use for all management communications with the system.

  So far, this is not too much of a problem. Many modern routers (and other devices) with many interfaces support management access via a specific interface/address or a selection of the interfaces available to the system. The current system, however, unlike smaller systems with a single system identity, is really a loose collection of computer systems in a network, each with a different identity. Given this architecture, most people would probably select only one or two system components to hold the consolidated management view. A unified view of the entire system must be synthesized on those system components that support the management function.

  The interfaces that we have been discussing are a perfect example of the need for consolidation of information/view. Whether accessing the system from a CLI or from SNMP at the management address, the user wants a view of all the operational interfaces on the device, plus their status and counter information—not just those local to the CPU (system) whose IP address the user provided for SNMP or CLI access. There is likely to be only one system component having this integrated view, meaning that only certain external interfaces on our system will respond to management requests. In some cases, the software having the information will be on the system component with interfaces to the management address. In others, the consolidated view (or data needed to create it) may be on another system component and the request may have to be forwarded via one of the internal interfaces to another CPU or system, processed, and then returned over the internal network back to the system that received the user request. The response is then sent back to the user. The problems associated with having a "private" set of internal addresses along with the externally visible ones affect much of the operational software in the system.*

- **Statistics and counter aggregation**—Users want a single aggregated view of information. They'll want to know, for example, the aggregate number of packets that have been forwarded by a system, or the total number of Web pages a system has served. In each case, many system components may have been involved in delivering the service. For a distributed system to provide this information, the management software will also have to be distributed throughout the system to collect and aggregate information. Then, when requests for the total number of packets forwarded on all the interfaces come in, they can be answered in an accurate and timely fashion.

- **Coordinated configuration**—When a management address receives a management request involving configuration, many system components may be affected. The management software that receives the request must have the intelligence to parcel out the configuration commands for execution on the proper system components. Because these system components often behave like separate computer systems, the parceling out can get pretty tricky. Also tricky is the creation of a single integrated configuration file that operators can view to get the whole picture. Once again, each of the systems is a separate entity, and the management software must be able to glue together the separate components of configuration on each system in the box into something cohesive.

- **Partial access to the system**—A system of the kind under discussion is large in terms of capacity and the number of different functions it can perform. Therefore, it must support access by a wider range of operational personnel in more organizations than might traditionally be the case. The management software that controls access must know these other parts of the system. It must also know which data that represent those parts are available for modification or viewing by different operators. The SNMPv3 framework can be particularly helpful with this task.

---

*A number of reasons cause vendors to take the approach of creating a separate private address space inside their devices. This separate space can help make the device look less complex from an external perspective, and it does add a bit of security.

* **Redundant components**—If any component fails, including a management component, the management system must still operate. In the case of an operational component, the operational software must discover the failure and switch over to the redundant component. At the same time, the management software must also be notified. The problem for the management software in this case is keeping track of counters on two different interfaces or Web servers that are providing the same service on behalf of a customer. The management software must also know which component is in active or standby mode.

  Depending on the level of accounting or reliability required, a backup may be needed for management system components as well. How such a backup affects the cost of management software, (and often of hardware), depends on how tightly synchronized the backup and primary systems are.

* **User views of the system**—In talking about a single IP address for management functions throughout the system, we recognized the desire to have all interfaces aggregated into a single view for the examination of status and counter information. Most users also actively desire a view showing all the subsystems that make up the system (e.g., the different system and interface cards in a very large router) and the state of each interface on each system, for times when failure occurs on a subcomponent, such as an interface. If a failure occurs on an interface, we may want a different view that shows all the boards in the system and the state of each interface on each. Other helpful and legitimate views would show services and customers that may span system components in this distributed device. Imagine a case where a service that guarantees low latency for a customer is deployed on a large system with many interfaces. From an operational perspective, one important view is the state of the interfaces that deliver this service for a particular customer. A failure of one of these interfaces might be handled differently from the way a failure on an interface that does not carry high-priority traffic. is handled Such a special view can help operational personnel zero in on important problems. Users of the management system will certainly want to select the view they want from their single management address. Indeed, setting it up any other way won't work: even if you require separate log-in to different addresses for different views, the views generally span system components.

Vendors can tackle each of these issues in a number of ways, but all operational personnel want to know is whether their vendor has addressed the issues in a way that makes the management task easier or harder.

Transactional Integrity

Although "out of the box" the SNMP framework supports only two basic levels of transaction control, it provides the mechanisms by which we can build any level necessary. Transaction control, however, is not transmission reliability. When we speak of transmission reliability, we are talking about whether the receiving device (usually the managed device in this context) can determine whether the data received is intact. Most protocols, including the predominantly UDP-based SNMP, have this property. When you send a MIB Object with a value of 5, you can be reasonably sure that a 5 was delivered. If the packet is corrupted during transmission, the receiving device detects the corruption and will not accept the information. Thus, transmission reliability ensures *information integrity*: the information we intended to send got to its destination in the form that we intended—it was reliably transmitted.

As important as transmission reliability, but quite distinct, is the property of transaction control. We'll get deeper into transactions and transaction control when we discuss management applications. For now, there are four levels of transaction control essential to managed systems that use SNMP for configuration, as depicted in Figure 8.8.

Transaction Level

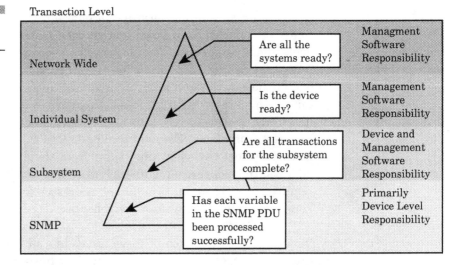

Figure 8.8
Transaction control.

※ **The network-wide configuration action**—For correct service activation and change, we may want to coordinate configuration actions across many machines. All the managed system needs to meet this requirement is a *handle*, which is the big switch for a transaction that may contain many individual messages. SNMP provides several tools to give MIB designers the ability to create transaction control objects.

Transaction control is not the same as informational integrity, but transaction control has little value without information integrity. Beyond MIB Objects, a managed device doesn't have much responsibility for this level of transaction control, which is primarily the purview of the management station. It's the station that keeps track of all the managed systems that are part of a network-wide transaction and determines when all are ready for the global commit. A management system must be equipped for the possibility that one or more systems may fail during the final commit. If such a failure should occur, the management station consults its configuration information and either rolls back the entire transaction across all the managed devices or determines that the failures are OK and simply logs them.

※ **The "local transactions complete" level**—Conceptually, this level might be merged with the previous one. Indeed, the previous level is the sum of all the individual local transaction-complete results for each system in the scope of a specific network-wide configuration transaction. We must provide the managed system's ability to accept a large number of smaller operations that may configure many different subparts of the system. Configuration of the routing and QoS mechanisms are examples of subsystems in a managed device that may require many smaller transactions (see the next section on subsystem completion) before each subsystem has all of the information needed for a single larger transaction (reconfiguration) of that subsystem. At the completion of all of these subsystem configuration setups, we want all the system configuration changes to have been received and be good to go.

The management system is likely to be the only entity that "knows" about all the configuration actions required for a transaction to be completed on each managed system. It may have to evaluate differences from one machine to the next to determine completeness. The management station will most likely be where rollback and other error recovery mechanisms take place. At the local device-complete level, a managed system may report configuration errors and perform some level of error correction on its own. If the management station has left a subsystem unconfigured or incorrectly configured, the managed system should report the error by this stage or at least before the final commit is issued.

- **Subsystem completion**—Whether you need to configure many different aspects of a system or just one, the same technique can be used to activate the configuration changes at the subsystem, system, or network-wide level. However, it is likely, and even desirable, that there will be separate management objects that activate subsystem and global configuration changes on a managed device. This gives the management application several granularities to use, depending on different needs, which is helpful even though the managed device may not have much understanding of the network-wide level—in fact, it does not need to, as long as the management software does. While the managed device will be able to send a notification if an error in subsystem configuration exists, the error may not be reported until the final commit is issued. The management station must always be aware of each subsystem's requirements for correct configuration on each managed device. Without this awareness, the odds of a correct configuration operation are low. If the management station doesn't possess the proper awareness, the person running the management application will have to know all the details (just as they do today when performing CLI actions or writing scripts for CLI-based systems). This approach isn't scalable and isn't suited for an effective service management system.

- **SNMP level completion**—This level is built into the base SNMP facilities. At present, a transaction can be thought of as confined to a single SNMP Protocol Data Unit. That is, there is no state in the SNMP system from one protocol data unit to the next. Of course the management application does get a confirmation of success for each SetRequest-PDU. If a single variable within a SetRequest-PDU fails on a managed device, an error can be sent back to the manager with meaningful information about the failure. In a well-designed system, the management station takes the error information and retries (if possible) based on this error and perhaps other information. If the error is not really an SNMP protocol error, the system may send a notification message back to the managed device containing additional information useful to the manager in its retry attempt.

  While this simple approach does have advantages in terms of efficiency and ease of implementation, it does not provide the scope needed to cover very large (re)configuration activities, where we may have much more information to send to a managed system than would fit into a single PDU. Is this as serious a problem with SNMP as some have claimed? I don't believe so. In fact, even connection-oriented protocols often have to resort to sending multiple messages, and sometimes messages of different types, over a TCP connection to carry all the information necessary. If the TCP connection is broken, state information may be lost and data transmission may have to begin again. When using SNMP, on the other hand, if we lose a configuration message (SNMP PDU) all we have to do is retransmit that single PDU to those devices that did not acknowledge receipt. We never have to send all messages over again.

To focus on the size issue too much is to miss an important distinction about transaction control: even if we were to have the ability to reliably communicate arbitrarily large amounts of data to a system over TCP or with a very large UDP packet, we still need transaction-level semantics to communicate transaction start, stop, and other operations. Using the SNMP-based approach, we can use MIB Objects to signal a commit for one or many MIB Objects whose values have been (re)configured during a transaction. In short, we can use MIB Objects for transaction semantics at any level of transaction granularity we need. Some might argue such transaction semantics should be built into the protocol. But where do you draw the line in terms of how much complexity and function to build into the protocol? In the end, only the application really knows what a transaction is.

Management Software Meets Operational Software

How well the management system functions is largely dependent on how well the operational software has been designed and implemented to meet the needs of the management software. As a general rule, most vendors place greatest emphasis on achieving basic operational functions for their products, and only then address the needs of the management software. In this section, we examine some of the design issues that get short shrift when vendors focus primarily on the operational software.

Transactional Support in the Operational Software

Earlier, we defined a number of types of transactional control and granularity. Without support in the operational software, it is difficult for the management system to provide effectively for all the different levels of transaction needed. Operational software can also be limited to some degree by the hardware on which it runs, so, in a sense, device manageability is influenced by a series of dependencies.

▓ **Granularity of changes for system or subsystem (re)initialization**—A system or subsystem may require the configuration of many individual parameters to operate correctly. The granularity supported by the operational software amounts to how many of them must be changed as a group. At one end of the spectrum are systems with essentially no granularity; these must be reinitialized and all parameters set simultaneously for even a single parameter change. The classic example is the system that must read an entire configuration file to implement the most minor revision on one of potentially many lines within that configuration file. Most modern systems are somewhat more flexible than this. At the other extreme are systems that permit any level of granularity for configuration changes. If even a single configuration parameter is sent to the operational code, it accepts it without requiring reinitialization of the entire system. Of course, this kind of flexibility is more costly to implement and may not always make sense from the perspective of the operational software.

Developers of operational code are sometimes bound by the standards requirements of the operational code. This constraint has two dimensions. First, the protocol specification identifies how a particular protocol is to behave. In some cases, the specifications don't permit the change of certain parameters without subsystem reinitialization, and in other cases they do. Therefore operational code developers are not the only folks who need to understand the protocol specifications; everyone building the management infrastructure, up to and including the management application developers, must also know how the protocol works.

The second standards requirement for operational code arises from de facto conventions that have come about as the result of the dominance of one vendor or another. From a practical perspective, the distinction between these two requirements is not so important. What does matter is correct behavior in the network, and formal standards are of less importance than system interoperability across vendors. How rapidly a system or subsystem can be reinitialized is a consideration you should factor into your design choices for management software. Should you allow or disallow fine-grained changes (changes to very small parts of a system or subsystem), or are you prepared to allow them only under certain circumstances, when reinitialization times are not a significant concern?

※ **System and subsystem initialization latency**—Initialization latency in the operation code is also impacted by standards. Many protocol specifications have timer and time-out values, so that if a system takes too long to reinitialize, the network won't think it's operational. Note that initialization latency is measured inclusive of any latency introduced by the management software.

Sometimes the management software can offer a few tricks to help in this area by preprocessing changes in systems that have this facility. When it comes time to put the new changes online, they can be predigested so that the system doesn't have to start from scratch. One way to enable preprocessing is to let an off-line version of the protocol interpret the new configuration parameters. The management system can then bring the protocol on line when needed.

※ **Rollback**—In some cases, a configuration change results in unanticipated side effects, necessitating a rollback to a previous version of the configuration. Just how easy it will be to perform the rollback depends on granularity and latency, as well as the operational code's flexibility to maintain separate versions of the configuration data. In a system where the management software and the operational software are tightly integrated, the operational software may not know about different versions, and must be given a specific file or pointer to new configuration data the management subsystem.

The considerations just listed are so closely related that they must be evaluated as a whole. For example, if a subsystem takes a relatively long time to initialize even for small granularity changes, and if this latency presents significant impacts to the network, you're going to want to be very sparing in the number of changes permitted to trigger a subsystem initialization. If, on the other hand, a system reinitializes very rapidly and without serious side effects on network behavior, it might be feasible to reconfigure it more frequently. This could yield you shorter time lag for new service deployments and error-correcting configuration changes.

Well-designed management software can sometime mask deficiencies in the operational code. In a system that does support fine-grained changes, but takes a long time to reinitialize, you may want a management infrastructure that stores changes on the local device for simultaneous application. Using this approach, transactions for all the individual changes are held in a "pending" state until a selected time when all the incremental changes take effect. This kind of management system design provides reasonable operational flexibility without having an adverse impact on the stability of the network. Pending transactions can be held on the managed device simply by creating MIB Objects to control the application of configuration changes.

It makes little difference whether the method by which the configuration information is sent to the device is CLI, SNMP, or something else. Transaction support requirements in the operational software are always the same. The sole difference between methods is the way management information and operations are converted into commands that the operational software can understand. The structure of the operational software may have an effect on CLIs, just as it will have an impact on MIB Modules that are designed for a particular system.

Dealing with Counters

When they're developing a management infrastructure in a managed device, engineers often ask how often counters found in MIB Objects should be updated, and how these updates should be accomplished. In practice, different counters have different requirements and you have a range of options.

Earlier, we discussed the memory and CPU implications of data collection for counter objects. Our current context is SNMP counter objects, but the essentials are the same regardless of how counter information is transferred to a management system.

Figure 8.9 illustrates two mechanisms by which SNMP counter objects can obtain the low-level information they need to be able to respond to SNMP requests for information. In the first method, the SNMP agent requests information from the low-level operating software. For example, *ifInOctets* [RFC2863] might access raw data in the operating software on a periodic basis. It makes the request using whatever native facilities available for moving information between different parts of the system, usually the internal messaging system. Once the information is collected and perhaps synthesized with other information, the counter is available for retrieval by any manager sending SNMP requests for information (e.g., GetBulk) [RFC1905].

The second mechanism that SNMP agents can use to realize a MIB Object is to receive periodic update information from the operational code. In this case, the operational code sends out information to the system about counters. This approach is commonly used for notifications as well.

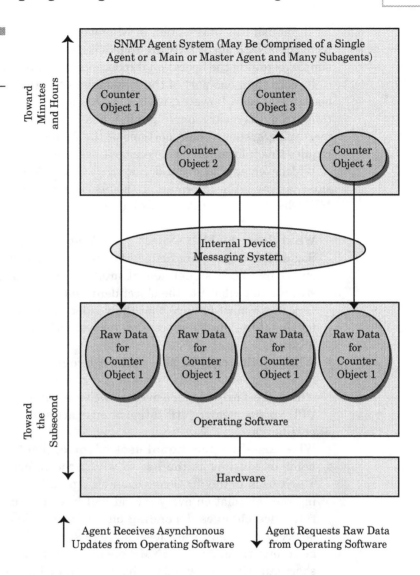

Figure 8.9
Counters inside a
managed device.

The decision to have the SNMP code collect data or have the operational code supply counter data is usually based on which other parts of the system need the information. If the counter data is primarily used by the SNMP system, then the burden of collection might be placed on the SNMP agent code. Thus, the SNMP agent collects the raw counter data through whatever API is available in the operational code. The decision is influenced by facilities in the operation code: where operational code does not have a convenient data interface, it may be better to let it publish this information in whatever form is convenient for it, and have the SNMP agent software collect it. In some cases, this might be as simple as telling the operational code to periodically write to a plain file.

If more than one part of the system uses the raw counter data in the operational code, however, it often makes more sense to have the operational code asynchronously publish the information over the internal messaging system. Those portions of the managed device software that require this information will receive it over the asynchronous interface.

Whichever approach you decide to use, an important question to consider is how often information should be updated for reflection in the MIB Objects. When making this judgment, consider the following points:

- **What does the MIB Object specification require (if anything)?** Some MIB Objects are defined for update every N seconds. When an object has been so defined, it must be updated within these constraints to conform to the object definition. For a vendor-specific MIB Object, we must assume that the vendor has carefully considered the impact of the update process on its system and chosen a wise update interval.
- **What is the cost of the update process?** With respect to CPU resources, some counter objects are costly to make available while others are fairly inexpensive. As we've mentioned, improvements in CPU versus memory utilization often exact a trade-off, and this is one possible case.
- **What are the operational uses of the object?** The most important point to consider is the use to which the information will be put. Think back to the ifInOctets example again. It is unlikely that updating the information every second will have real utility, but the time reference changes depending on where you look in the managed device. At the hardware level on high-speed interfaces, a second can be a long time. At the SNMP agent level, a second can be a fairly short duration. So, while the hardware may have some registers and counters that are eventually exposed as SNMP objects, that does not imply that the system should go through the expense of updating SNMP counters as frequently as other parts of the system. The SNMP agent can collect data, keep internal track, and then update the value for a MIB Object on a frequency that makes sense, given the definition or intended use of that object.

These approaches assume a model in which the counter information is always up to date and available when the request is received without too much additional computation by the subagent. A different approach, suited to objects that are not frequently used, is for the SNMP agent to request information only when it needs it to respond to an SNMP request. Take care in creating this kind of object, because these objects tend to cause spikes in the system load. If a lot of such SNMP requests come in at once, the overall performance of the system could be adversely affected. This approach will not work well where the objects require counters over specific, or potentially long periods of time, and are quite complex. A particular problem with counters that must hold data over long periods is that the operational software or the hardware may not have the facilities to store the information until requested by the SNMP agent. The DS1/E1/DS2/E2 MIB Module [RFC2495] has several examples of time-specific objects. How they're implemented in a particular system is determined in part by the architecture of that system.

Two terms used by objects in the DS1/E1/DS2/E2 MIB describe the complexity: Severely Errored Framing Second (SEFS) and Unavailable Seconds (UAS). These terms are defined in RFC 2495 that follows:

```
Severely Errored Framing Second (SEFS)
        An Severely Errored Framing Second is a second with one
        or more Out of Frame defects OR a detected AIS defect.
        (Also known asSAS-P (SEF/AIS second); See T1.231 Section
        6.5.2.6)

    dsx1CurrentSEFSs OBJECT-TYPE
        SYNTAX    PerfCurrentCount
        MAX-ACCESS    read-only
        STATUS    current
        DESCRIPTION
            "The number of Severely Errored Framing Seconds."
        ::= { dsx1CurrentEntry 4 }

Unavailable Seconds (UAS)
        Unavailable Seconds (UAS) are calculated by counting the
        number of seconds that the interface is unavailable.  The
        DS1 interface is said to be unavailable from the onset of
        10 contiguous SESs, or the onset of the condition leading
        to a failure (see Failure States). If the condition
        leading to the failure was immediately preceded by one or
        more contiguous SESs, then the DS1 interface
        unavailability starts from the onset of these SESs.  Once
        unavailable, and if no failure is present, the DS1
        interface becomes available at the onset of 10 contiguous
        seconds with no SESs.  Once unavailable, and if a failure
        is present, the DS1 interface becomes available at the
        onset of 10 contiguous seconds with no SESs, if the
        failure clearing time is less than or equal to 10
        seconds.  If the failure clearing time is more than 10
```

seconds, the DS1 interface becomes available at the onset
of 10 contiguous seconds with no SESs, or the onset
period leading to the successful clearing condition,
whichever occurs later. With respect to the DS1 error
counts, all counters are incremented while the DS1
interface is deemed available. While the interface is
deemed unavailable, the only count that is incremented is
UASs.

Note that this definition implies that the agent cannot
determine until after a ten second interval has passed
whether a given one-second interval belongs to available
or unavailable time. If the agent chooses to update the
various performance statistics in real time then it must
be prepared to retroactively reduce the ES, BES, SES, and
SEFS counts by 10 and increase the UAS count by 10 when
it determines that available time has been entered. It
must also be prepared to adjust the PCV count and the DM
count as necessary since these parameters are not
accumulated during unavailable time. It must be
similarly prepared to retroactively decrease the UAS
count by 10 and increase the ES, BES, and DM counts as
necessary upon entering available time. A special case
exists when the 10 second period leading to available or
unavailable time crosses a 900 second statistics window
boundary, as the foregoing description implies that the
ES, BES, SES, SEFS, DM, and UAS counts the PREVIOUS
interval must be adjusted. In this case successive GETs
of the affected dsx1IntervalSESs and dsx1IntervalUASs
objects will return differing values if the first GET
occurs during the first few seconds of the window.

The agent may instead choose to delay updates to the
various statistics by 10 seconds in order to avoid
retroactive adjustments to the counters. A way to do
this is sketched in Appendix B."

```
dsx1CurrentUASs OBJECT-TYPE
    SYNTAX  PerfCurrentCount
    MAX-ACCESS  read-only
    STATUS  current
    DESCRIPTION
          "The number of Unavailable Seconds."
    ::= { dsx1CurrentEntry 5 }
```

These objects have quite a lot of temporal context, as we can see from
the definitions that precede them. When using objects of this type, we
will most often have to perform calculations on an ongoing basis. Therefore, an approach that only gets an answer when a SNMP request comes
in might not work as well.

Dealing with Notifications

Notifications present many of the same concerns as counters, and are known to add to management problems as much as alleviate them. There are two main problems with the notification software often found in managed devices today: too many and too frequent.

Too many. Question: if a four-port Ethernet card fails, should the managed system send one, four, or five notifications? Answer: a single notification that the card has failed provides all the information that the other two alternatives would have provided. If five notifications were sent, one would have reported each port failure and one would have reported the card. It's unnecessary but not too harmful. In the case of four notifications, one would have reported each port failure and no notification would have been sent for the card. This is the potentially problematic example. It is possible, although not likely, that the card would still be functional with all four of its ports down. Of course a four-port Ethernet card with all ports in a failed state cannot perform much useful work, but the single notification approach is still preferable. It can only be done, however, when the management software in the device is smart enough to know the physical characteristics of the device. That is, the software sending notifications from the managed device must have access to information about ports and their relationship to cards, and must understand the card's relationship to the rest of the system. Given this information, the management software can make more intelligent decisions about whether to send a notification. We will discuss this scenario again in Chapter 10, *Specialized Function Subagents*.

In the example objects below from RFC 2863, you'll find reasonably well-defined MIB objects, but no guidance about what to do in case of a system where the interfaces are moving between states as a result of some failure in the managed device. Sometimes we can handle this case with better definitions. For now, we are left with a system having the potential to send too many notifications:

```
linkDown notification-TYPE
    OBJECTS { ifIndex, ifAdminStatus, ifOperStatus }
    STATUS  current
    DESCRIPTION
            "A linkDown trap signifies that the SNMP entity,
            acting in an agent role, has detected that the
            ifOperStatus object for one of its communication
            links is about to enter the down state from some
            other state (but not from the notPresent state).
```

```
                 This other state is indicated by the included value
                 of ifOperStatus."
::= { snmpTraps 3 }

ifOperStatus OBJECT-TYPE
    SYNTAX  INTEGER {
                 up(1),          -- ready to pass packets
                 down(2),
                 testing(3),    -- in some test mode
                 unknown(4),    -- status can not be determined
                                -- for some reason.
                 dormant(5),
                 notPresent(6),    -- some component is missing
                 lowerLayerDown(7) -- down due to state of
                                   -- lower-layer interface(s)
             }
    MAX-ACCESS  read-only
    STATUS      current
    DESCRIPTION
             "The current operational state of the interface.  The
             testing(3) state indicates that no operational
             packets can be passed.  If ifAdminStatus is down(2)
             then ifOperStatus should be down(2).  If
             ifAdminStatus is changed to up(1) then ifOperStatus
             should change to up(1) if the interface is ready to
             transmit and receive network traffic; it should
             change to dormant(5) if the interface is waiting for
             external actions (such as a serial line waiting for
             an incoming connection); it should remain in the
             down(2) state if and only if there is a fault that
             prevents it from going to the up(1) state; it should
             remain in the notPresent(6) state if the interface
             has missing (typically, hardware) components."
             ::= { ifEntry 8 }

ifAdminStatus OBJECT-TYPE
    SYNTAX  INTEGER {
                 up(1),          -- ready to pass packets
                 down(2),
                 testing(3)     -- in some test mode
             }
    MAX-ACCESS  read-write
    STATUS      current
    DESCRIPTION
             "The desired state of the interface.  The testing(3)
             state indicates that no operational packets can be
             passed.  When a managed system initializes, all
             interfaces start with ifAdminStatus in the down(2)
             state.  As a result of either explicit management
             action or per configuration information retained by
             the managed system, ifAdminStatus is then changed to
             either the up(1) or testing(3) states (or remains in
             the down(2) state)."
::= { ifEntry 7 }
```

Too frequent.   Using the objects given, an interface that is moving between states might threaten to emit many notifications in a small period of time without added benefit to operational personnel. There is no information in any of the notification or OBJECT-TYPE definitions that can help us rate limit notification generation. The problem can be solved in two ways: via better object and notification definitions and via specialized agents to process information in the managed devices more intelligently. We'll take up these topics again in the next two chapters. Right now, our focus is on the overall system design. To address the problem with the objects in our current discussion, we might create an agent that used another object as a throttle for overly frequent notifications. For example, the following object could be evaluated by notification software to help it determine if it wanted to send another notification:

```
ifLastChange OBJECT-TYPE
    SYNTAX      TimeTicks
    MAX-ACCESS  read-only
    STATUS      current
    DESCRIPTION
            "The value of sysUpTime at the time the interface
            entered its current operational state.  If the
            current state was entered prior to the last re-
            initialization of the local network management
            subsystem, then this object contains a zero value."
    ::= { ifEntry 9 }
```

To make this work, we could create agent software that monitored the value of ifLastChange. We could specify that the system will not send additional notifications more frequently than some elapsed time from the last ifLastChange. The problem with this solution is that we may want to be reminded each hour if the interface is still flapping moving between states—and for that we may need control objects to stipulate how long to hold off sending notifications.

The problem of "too many and too frequent" is so serious from an operations perspective that a number of vendors have developed tools to filter out extra notifications and forward only the most significant to operator management consoles. Advanced versions of these tools claim to perform root-cause analysis to determine the original cause of the failure. Many of these software products do add significant value.

The fact that an industry has sprung up, at least in part, in response to overly simplistic software management devices should not deter us from building software that generates more meaningful notifications.

The more effective the notifications, the better our third-party software will become at getting to the root cause of failures and reducing mean time to repair, an important consideration for anyone selling high-value managed services.

There may, in fact, be a relationship between how often a system is configured to generate notifications and the requirements of service-level agreements. In this discussion, we imagined creating control objects to allow us to adjust the frequency of certain notifications. Where service level agreements are in force, I can envision setting the value of various control objects for network resources (interfaces, Web servers, and application servers), to values that will fulfill customer agreements. A positive outcome of doing so is that we would then be generating notifications based on business rules instead of hard-coded values that may not reflect what matters to the network operator and the customers served. Here's another example of the powerful relationship between configuration and fault/performance data: a management system that knows about the desired configuration of network devices can also configure notification controls. It can process received notifications more intelligently, because it will have already set up the context in which to evaluate them. Good management software design, well-architected MIB Modules, and correct application of principles exemplified by the SNMPCONF technology in managed devices are prerequisites to building effective service management systems, but by themselves, these elements do not guarantee success. The next step is to find techniques for leveraging what we have thus far described for greater effectiveness. Although the approaches we discuss are presented in the context of SNMP-based technology, they are basic enough to apply to all kinds of successful management software. If you plan to use other technologies in managed devices, you can measure them against the functions described here to decide if they will be sufficient for your needs.

CHAPTER 9

SNMP and SNMPCONF Enable More than Configuration

SNMP and SNMPCONF* technology can be used for far more than configuration. We've already shown how state and utilization can be used by *policyCondition* and *policyAction* scripts to determine which elements to configure and the details of configuration operations. Effective configuration depends on many types of information about a device, in the same way that fault management depends on configuration information for most effective operation.

Configuration is only half the story for a service management system. This chapter demonstrates why monitoring the result of configuration activity is just as important as performing that activity correctly in the first place. It also gives readers some standard MIB Objects that are available for use in their own systems to monitor services that have been created by the configuration operations. More important than the specific MIB objects presented are the principles they are used to help illustrate.

Service-level Reporting

Imagine we have a shared resource like an application or a Web server, and we want it to run some services at very high levels of reliability because some of our customers have the right to exact financial penalties for a failures exceeding a set number of minutes per month. They may also be able to exact a penalty if users have to wait too long to see Web pages in exchange for the higher premium they pay.

Figure 9.1 depicts a Web or application server running a number of processes to support different service levels. We will use this fictional server to help deliver our high-value Web services. Some customers pay a premium for low-latency access to pages on this server. In the business model of this service provider, there are two basic types of services: those that support premier users, and those that support users who are only guaranteed basic or best-effort service. This particular server has been configured with two network interface cards, which, in turn, are configured to support preferential treatment for packets that are on their way to or from the premier users. As an added benefit, the multiple interfaces can be set up to connect to different network devices, thereby reducing the likelihood of a singe device bringing down the service.

---

*Although I describe SNMPCONF technology, it is not the specifics of SNMPCONF that matter as much as the principles that it incorporates. Specifically, you want to configure your devices in support of services and business policies. You want to manage the elements of your system not only as discrete elements, but also as pieces that support an integrated service using a common name space for configuration and all the other aspects of management.

Figure 9.1
Service-level
management with
SNMP and
SNMPCONF.

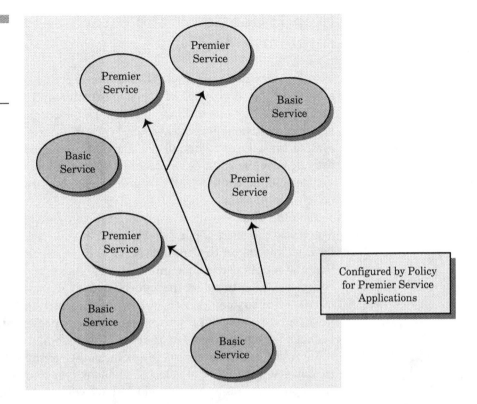

To feed our service management system enough data to determine whether service levels have been met, at least at a given network device, we must perform three functions without fail:

* Configure the network device to support the desired services for each customer
* Monitor the status of critical services in the network device
* Report successful and unsuccessful service delivery

The service management system will perform some important functions with regard to all of these tasks, but here we're only going to examine the role of the managed device.

Step 1: Configure the Interfaces and Applications

The first step in our example is to correctly configure the device to deliver each of the two types of service: premier and basic. This might be accomplished by one or many policies. In an upcoming chapter on management software, we discuss the types of information a management system must have to install policies on a managed device, but for now, assume that the right decisions have been made.

The task of the device is to implement, monitor, and report on the policies in question.

Assume that we have two policies relevant to the configuration aspects of the each of our two types of service, (premier and basic). For now, let's focus on the premier service. Each service has a *policyCondition* and a *policyAction* for interface configuration. The first policy has a *policyCondition* to select the interfaces that we wish to manage for premier service customers. For each of the interfaces selected, the *policyAction* configures specific parameters to give preferential treatment to packets determined to be on their way to or from premier users.

Interface selection is made by examining of the roles that have been assigned to interfaces of selected element types such as Fast Ethernet. Roles such as "premier" or "gold," along with a specific customer name assigned by management software to all interfaces used by that customer, would meet our needs. Keep in mind that different types of interfaces demand different parameters to cause the system to deliver the service level desired. Management software is responsible for sending the correct policies to each managed device, based on the element types supported in the device and found in the element type registration table (discussed in Chapter 7). Of course, managed devices should accept policies for element types they do not support and still behave correctly.* In this event, the system finds no elements under the control of the policy, because no element types match the *policyCondition*—a fact that the management station discovers by investigating the *pmPolicyMatches* object. If this is the case, this system could not deliver the services we want. For now, let's assume that at least a couple of interfaces do meet our requirements.

The second policy for our premier customers also consists of a *policyCondition* and a *policyAction*, but this policy is used to configure the

---

*Note that element type information is not the only information that we use to determine which policies are sent to managed devices. The information in the capabilities tables, along with other data, is used to help in this selection.

intended premier service processes on our application or Web server. The *policyCondition* ascertains that the desired processes exist and, if they do, the *policyAction* action ensures that they're correctly configured. In the absence of a specific process with identifiable parameters (such as process priority or a name), the *policyAction* starts the process with the correct parameters. This ensures that a specified Web service is run for our premier customers. This demonstrates how the managed system monitors some of its own configuration and subsequent behavior and takes corrective steps when things go wrong.

Step 2: Service-level Monitoring

Many objects that control at least some of the configuration parameters are proprietary to a specific vendor. On the other hand, many objects for service-level reporting and monitoring have been defined in the IETF for a wide array of element types and are available in popular products. This makes the task of writing policies for service-level monitoring and reporting somewhat simpler than its configuration-level counterpart.

In our application, we could use Definitions of System-Level Managed Objects for Applications [RFC2287] to determine if, and in what state, the processes are running. Two tables of particular interest are the *sysApplRunTable* and the *sysApplPastRunTables*, which give a top-level view of the state of running applications and those that have recently existed. These tables are reproduced here in full, for those readers who already work comfortably with MIB Objects. If you're not one of them, you'll make some sense of the MIB by reading only the paragraph(s) that follow the DESCRIPTION portion of the definition of each object.

```
-- sysApplRunTable
-- The sysApplRunTable contains the application instances
-- which are currently running on the host.  Since a single
-- application might be invoked multiple times, an entry is
-- added to this table for each INVOCATION of an application.
-- The table is indexed by sysApplInstallPkgIndex,
-- sysApplRunIndex to enable managers to easily locate all
-- invocations of a particular application package.

sysApplRunTable OBJECT-TYPE
    SYNTAX        SEQUENCE OF SysApplRunEntry
    MAX-ACCESS    not-accessible
    STATUS        current
    DESCRIPTION
        "The table describes the applications which are
         executing on the host.  Each time an application is
```

invoked, an entry is created in this table. When an
application ends, the entry is removed from this table
and a corresponding entry is created in the
SysApplPastRunTable.

A new entry is created in this table whenever the
agent implementation detects a new running process
that is an installed application element whose
sysApplInstallElmtRole designates it as being the
application's primary executable
(sysApplInstallElmtRole = primary(2)).

The table is indexed by sysApplInstallPkgIndex,
sysApplRunIndex to enable managers to easily locate
all invocations of a particular application package."
::= { sysApplRun 1 }

```
sysApplRunEntry OBJECT-TYPE
    SYNTAX       SysApplRunEntry
    MAX-ACCESS   not-accessible
    STATUS       current
    DESCRIPTION
        "The logical row describing an application which is
        currently running on this host."
    INDEX    { sysApplInstallPkgIndex, sysApplRunIndex }
    ::= { sysApplRunTable   1 }

SysApplRunEntry ::= SEQUENCE {
    sysApplRunIndex                     Unsigned32,
    sysApplRunStarted                   DateAndTime,
    sysApplRunCurrentState              RunState
}

sysApplRunIndex OBJECT-TYPE
    SYNTAX       Unsigned32 (1..'ffffffff'h)
    MAX-ACCESS   not-accessible
    STATUS       current
    DESCRIPTION
        "Part of the index for this table. An arbitrary
        integer used only for indexing purposes. Generally
        monotonically increasing from 1 as new applications
        are started on the host, it uniquely identifies
        application  invocations.
```

The numbering for this index increases by 1 for each
INVOCATION of an application, regardless of which
installed application package this entry represents a
running instance of.

An example of the indexing for a couple of entries is
shown below.

:

sysApplRunStarted.17.14

```
                        sysApplRunStarted.17.63
                        sysApplRunStarted.18.13
                               :
```

In this example, the agent has observed 12 application
invocations when the application represented by entry
18 in the sysApplInstallPkgTable is invoked. The next
invocation detected by the agent is an invocation of
installed application package 17. Some time later,
installed application 17 is invoked a second time.

NOTE: this index is not intended to reflect a real-
time (wall clock time) ordering of application
invocations; it is merely intended to uniquely
identify running instances of applications. Although
the sysApplInstallPkgIndex is included in the INDEX
clause for this table, it serves only to ease
searching of this table by installed application and
does not contribute to uniquely identifying table
entries."

```
    ::= { sysApplRunEntry 1 }

sysApplRunStarted OBJECT-TYPE
    SYNTAX      DateAndTime
    MAX-ACCESS  read-only
    STATUS      current
    DESCRIPTION
        "The date and time that the application was started."
    ::= { sysApplRunEntry 2 }

sysApplRunCurrentState OBJECT-TYPE
    SYNTAX      RunState
    MAX-ACCESS  read-only
    STATUS      current
    DESCRIPTION
        "The current state of the running application
        instance. The possible values are running(1),
        runnable(2) but waiting for a resource such as CPU,
        waiting(3) for an event, exiting(4), or other(5). This
        value is based on an evaluation of the running
        elements of this application instance (see
        sysApplElmRunState) and their Roles as defined by
        sysApplInstallElmtRole.  An agent implementation may
        detect that an application instance is in the process
        of exiting if one or more of its REQUIRED elements are
        no longer running.  Most agent implementations will
        wait until a second internal poll has been completed
        to give the system time to start REQUIRED elements
        before marking the application instance as exiting."
    ::= { sysApplRunEntry 3 }

    -- sysApplPastRunTable
    -- The sysApplPastRunTable provides a history of applications
    -- previously run on the host computer. Entries are removed
```

```
-- from the sysApplRunTable and corresponding entries are
-- added to this table when an application becomes inactive.
-- Entries remain in this table until they are aged out when
-- either the table size reaches a maximum as determined by
-- the sysApplPastRunMaxRows, or when an entry has aged to
-- exceed a time limit as set by sysApplPastRunTblTimeLimit.
--
-- When aging out entries, the oldest entry, as determined by
-- the value of sysApplPastRunTimeEnded, will be removed
-- first.
```

```
sysApplPastRunTable OBJECT-TYPE
        SYNTAX          SEQUENCE OF SysApplPastRunEntry
        MAX-ACCESS  not-accessible
        STATUS          current
        DESCRIPTION
            "A history of the applications that have previously
            run on the host computer.  An entry's information is
            moved to this table from the sysApplRunTable when the
            invoked application represented by the entry ceases to
            be running.

            An agent implementation can determine that an
            application invocation is no longer running by
            evaluating the running elements of the application
            instance and their Roles as defined by
            sysApplInstallElmtRole.  Obviously, if there are no
            running elements for the application instance, then
            the application invocation is no longer running. If
            any one of the REQUIRED elements is not running, the
            application instance may be in the process of exiting.
            Most agent implementations will wait until a second
            internal poll has been completed to give the system
            time to either restart partial failures or to give all
            elements time to exit.  If, after the second poll,
            there are REQUIRED elements that are not running, then
            the application instance may be considered by the
            agent implementation to no longer be running.

            Entries remain in the sysApplPastRunTable until they
            are aged out when either the table size reaches a
            maximum as determined by the sysApplPastRunMaxRows, or
            when an entry has aged to exceed a time limit as set
            by sysApplPastRunTblTimeLimit.

            Entries in this table are indexed by
            sysApplInstallPkgIndex, sysApplPastRunIndex to
            facilitate retrieval of all past run invocations of a
            particular installed application."
        ::= { sysApplRun 2 }

sysApplPastRunEntry OBJECT-TYPE
        SYNTAX          SysApplPastRunEntry
        MAX-ACCESS  not-accessible
```

```
            STATUS       current
            DESCRIPTION
                "The logical row describing an invocation of an
                application which was previously run and has
                terminated.  The entry is basically copied from the
                sysApplRunTable when the application instance
                terminates.  Hence, the entry's value for
                sysApplPastRunIndex is the same as its value was for
                sysApplRunIndex."
            INDEX    { sysApplInstallPkgIndex, sysApplPastRunIndex }
            ::= { sysApplPastRunTable    1 }

        SysApplPastRunEntry ::= SEQUENCE {
            sysApplPastRunIndex                    Unsigned32,
            sysApplPastRunStarted                  DateAndTime,
            sysApplPastRunExitState                INTEGER,
            sysApplPastRunTimeEnded                DateAndTime
        }

        sysApplPastRunIndex OBJECT-TYPE
            SYNTAX       Unsigned32 (1..'ffffffff'h)
            MAX-ACCESS   not-accessible
            STATUS       current
            DESCRIPTION
                "Part of the index for this table. An integer
                matching the value of the removed sysApplRunIndex
                corresponding to this row."
            ::= { sysApplPastRunEntry 1 }

        sysApplPastRunStarted OBJECT-TYPE
            SYNTAX       DateAndTime
            MAX-ACCESS   read-only
            STATUS       current
            DESCRIPTION
                "The date and time that the application was started."
            ::= { sysApplPastRunEntry 2 }

        sysApplPastRunExitState OBJECT-TYPE
            SYNTAX       INTEGER {
                         complete (1), -- normal exit at sysApplRunTimeEnded
                         failed (2),  -- abnormal exit
                         other (3)
                         }
            MAX-ACCESS   read-only
            STATUS       current
            DESCRIPTION
                "The state of the application instance when it
                terminated. This value is based on an evaluation of the
                running elements of an application and their Roles as
                defined by sysApplInstallElmtRole.  An application
                instance is said to have exited in a COMPLETE state and
                its entry is removed from the sysApplRunTable and added
                to the sysApplPastRunTable when the agent detects that
```

```
            ALL elements of an application invocation are no longer
            running.  Most agent implementations will wait until a
            second internal poll has been completed to give the
            system time to either restart partial failures or to
            give all elements time to exit.  A failed state occurs
            if, after the second poll, any elements continue to run
            but one or more of the REQUIRED elements are no longer
            running. All other combinations MUST be defined as
            OTHER."
        ::= { sysApplPastRunEntry 3 }

    sysApplPastRunTimeEnded OBJECT-TYPE
        SYNTAX        DateAndTime
        MAX-ACCESS    read-only
        STATUS        current
        DESCRIPTION
            "The DateAndTime the application instance was
            determined to be no longer running."
    ::= { sysApplPastRunEntry 4 }
```

A close examination of these two tables may leave some readers wondering how to distinguish one application instance from another, given that not all applications are covered by the policy. [RFC2287] has tables that are associated with the two presented here and can help in this regard. These tables are the *sysApplElmtRunTable* and *sysApplElmtPastRunTable*. The *sysApplElmtRunInvocID* helps associate a specific part of an application, such as a process with a specific instance of an application. In our example, the Web server configured to serve our premier customers. Other objects in these tables tell us if the system is functioning correctly and how much of the hardware resources are being or have been used (in the case of the various "*pastRun*" tables):

```
    -- sysApplElmtRunTable
    -- The sysApplElmtRunTable contains an entry for each process
    -- that is currently running on the host.  An entry is created
    -- in this table for each process at the time it is started,
    -- and will remain in the table until the process terminates.
    --
    -- The table is indexed by sysApplElmtRunInstallPkg,
    -- sysApplElmtRunInvocID, and sysApplElmtRunIndex to make it
    -- easy to locate all running elements of a particular invoked
    -- application which has been installed on the system.


    sysApplElmtRunTable OBJECT-TYPE
        SYNTAX        SEQUENCE OF SysApplElmtRunEntry
        MAX-ACCESS    not-accessible
        STATUS        current
        DESCRIPTION
            "The table describes the processes which are currently
```

executing on the host system. Each entry represents a
running process and is associated with the invoked
application of which that process is a part, if
possible. This table contains an entry for every
process currently running on the system, regardless of
whether its 'parent' application can be determined.
So, for example, processes like 'ps' and 'grep' will
have entries though they are not associated with an
installed application package.

Because a running application may involve
more than one executable, it is possible to have
multiple entries in this table for each application.
Entries are removed from this table when the process
terminates.

The table is indexed by sysApplElmtRunInstallPkg,
sysApplElmtRunInvocID, and sysApplElmtRunIndex to
facilitate the retrieval of all running elements of a
particular invoked application which has been
installed on the system."
```
    ::= { sysApplRun 3 }


sysApplElmtRunEntry OBJECT-TYPE
    SYNTAX        SysApplElmtRunEntry
    MAX-ACCESS  not-accessible
    STATUS        current
    DESCRIPTION
        "The logical row describing a process currently
        running on this host.  When possible, the entry is
        associated with the invoked application of which it
        is a part."
    INDEX    { sysApplElmtRunInstallPkg,
sysApplElmtRunInvocID,
            sysApplElmtRunIndex }
    ::= { sysApplElmtRunTable   1 }


SysApplElmtRunEntry ::= SEQUENCE {
    sysApplElmtRunInstallPkg        Unsigned32,
    sysApplElmtRunInvocID           Unsigned32,
    sysApplElmtRunIndex             Unsigned32,
    sysApplElmtRunInstallID         Unsigned32,
    sysApplElmtRunTimeStarted       DateAndTime,
    sysApplElmtRunState             RunState,
    sysApplElmtRunName              LongUtf8String,
    sysApplElmtRunParameters        Utf8String,
    sysApplElmtRunCPU               TimeTicks,
    sysApplElmtRunMemory            Gauge32,
    sysApplElmtRunNumFiles          Gauge32,
    sysApplElmtRunUser              Utf8String
}


sysApplElmtRunInstallPkg OBJECT-TYPE
    SYNTAX        Unsigned32 (0..'ffffffff'h)
```

```
    MAX-ACCESS  not-accessible
    STATUS      current
    DESCRIPTION
        "Part of the index for this table, this value
        identifies the installed software package for
        the application of which this process is a part.
        Provided that the process's 'parent' application can
        be determined, the value of this object is the same
        value as the sysApplInstallPkgIndex for the
        entry in the sysApplInstallPkgTable that corresponds
        to the installed application of which this process
        is a part.

        If, however, the 'parent' application cannot be
        determined, (for example the process is not part
        of a particular installed application), the value
        for this object is then '0', signifying that this
        process cannot be related back to an application,
        and in turn, an installed software package."
    ::= { sysApplElmtRunEntry 1 }

sysApplElmtRunInvocID OBJECT-TYPE
    SYNTAX      Unsigned32 (0..'ffffffff'h)
    MAX-ACCESS  not-accessible
    STATUS      current
    DESCRIPTION
        "Part of the index for this table, this value
        identifies the invocation of an application of which
        this process is a part.  Provided that the 'parent'
        application can be determined, the value of this
        object is the same value as the sysApplRunIndex for
        the corresponding application invocation in the
        sysApplRunTable.

        If, however, the 'parent' application cannot be
        determined, the value for this object is then '0',
        signifying that this process cannot be related back
        to an invocation of an application in the
        sysApplRunTable."
    ::= { sysApplElmtRunEntry 2 }

sysApplElmtRunIndex OBJECT-TYPE
    SYNTAX      Unsigned32 (0..'ffffffff'h)
    MAX-ACCESS  not-accessible
    STATUS      current
    DESCRIPTION
        "Part of the index for this table.  A unique value
        for each process running on the host.  Wherever
        possible, this should be the system's native, unique
        identification number."
    ::= { sysApplElmtRunEntry 3 }


sysApplElmtRunInstallID OBJECT-TYPE
```

```
        SYNTAX      Unsigned32 (0..'ffffffff'h)
        MAX-ACCESS  read-only
        STATUS      current
        DESCRIPTION
            "The index into the sysApplInstallElmtTable. The
            value of this object is the same value as the
            sysApplInstallElmtIndex for the application element
            of which this entry represents a running instance.
            If this process cannot be associated with an installed
            executable, the value should be '0'."
        ::= { sysApplElmtRunEntry 4 }

sysApplElmtRunTimeStarted OBJECT-TYPE
        SYNTAX      DateAndTime
        MAX-ACCESS  read-only
        STATUS      current
        DESCRIPTION
            "The time the process was started."
        ::= { sysApplElmtRunEntry 5 }

sysApplElmtRunState OBJECT-TYPE
        SYNTAX      RunState
        MAX-ACCESS  read-only
        STATUS      current
        DESCRIPTION
            "The current state of the running process. The
            possible values are running(1), runnable(2) but
            waiting for a resource such as CPU, waiting(3) for an
            event, exiting(4), or other(5)."
        ::= { sysApplElmtRunEntry 6 }

sysApplElmtRunName OBJECT-TYPE
        SYNTAX      LongUtf8String
        MAX-ACCESS  read-only
        STATUS      current
        DESCRIPTION
            "The full path and filename of the process.
            For example, '/opt/MYYpkg/bin/myyproc' would
            be returned for process 'myyproc' whose execution
            path is '/opt/MYYpkg/bin/myyproc'."
        ::= { sysApplElmtRunEntry 7 }

sysApplElmtRunParameters OBJECT-TYPE
        SYNTAX      Utf8String
        MAX-ACCESS  read-only
        STATUS      current
        DESCRIPTION
            "The starting parameters for the process."
::= { sysApplElmtRunEntry 8 }

sysApplElmtRunCPU OBJECT-TYPE
        SYNTAX      TimeTicks
        MAX-ACCESS  read-only
        STATUS      current
```

```
DESCRIPTION
      "The number of centi-seconds of the total system's
      CPU resources consumed by this process.  Note that
      on a multi-processor system, this value may
      have been incremented by more than one centi-second
      in one centi-second of real (wall clock) time."
::= { sysApplElmtRunEntry 9 }

sysApplElmtRunMemory OBJECT-TYPE
      SYNTAX       Gauge32
      UNITS        "Kbytes"
      MAX-ACCESS   read-only
      STATUS       current
      DESCRIPTION
         "The total amount of real system memory measured in
         Kbytes currently allocated to this process."

      ::= { sysApplElmtRunEntry 10 }

sysApplElmtRunNumFiles OBJECT-TYPE
      SYNTAX       Gauge32
      MAX-ACCESS   read-only
      STATUS       current
      DESCRIPTION
         "The number of regular files currently open by the
         process.  Transport connections (sockets)
         should NOT be included in the calculation of
         this value, nor should operating system specific
         special file types."
      ::= { sysApplElmtRunEntry 11 }

sysApplElmtRunUser OBJECT-TYPE
      SYNTAX       Utf8String
      MAX-ACCESS   read-only
      STATUS       current
      DESCRIPTION
         "The process owner's login name (e.g. root)."
      ::= { sysApplElmtRunEntry 12 }

-- sysApplElmtPastRunTable
-- The sysApplElmtPastRunTable maintains a history of
-- processes which have previously executed on
-- the host as part of an application. Upon termination
-- of a process, the entry representing the process is removed
-- from the sysApplElmtRunTable and a corresponding entry is
-- created in this table provided that the process was part of
-- an identifiable application.  If the process could not be
-- associated with an invoked application, no corresponding
-- entry is created. Hence, whereas the sysApplElmtRunTable
-- contains an entry for every process currently executing on
-- the system, the sysApplElmtPastRunTable only contains
-- entries for processes that previously executed as part of
-- an invoked application.
--
```

```
-- Entries remain in this table until they are aged out when
-- either the number of entries in the table reaches a
-- maximum as determined by sysApplElmtPastRunMaxRows, or
-- when an entry has aged to exceed a time limit as set by
-- sysApplElmtPastRunTblTimeLimit.  When aging out entries,
-- the oldest entry, as determined by the value of
-- sysApplElmtPastRunTimeEnded, will be removed first.
--
-- The table is indexed by sysApplInstallPkgIndex (from the
-- sysApplInstallPkgTable), sysApplElmtPastRunInvocID, and
-- sysApplElmtPastRunIndex to make it easy to locate all
-- previously executed processes of a particular invoked
-- application that has been installed on the system.


sysApplElmtPastRunTable OBJECT-TYPE
    SYNTAX        SEQUENCE OF SysApplElmtPastRunEntry
    MAX-ACCESS    not-accessible
    STATUS        current
    DESCRIPTION
        "The table describes the processes which have
        previously executed on the host system as part of an
        application. Each entry represents a process which has
        previously executed and is associated with the invoked
        application of which it was a part.  Because an
        invoked application may involve more than one
        executable, it is possible to have multiple entries in
        this table for each application invocation. Entries
        are added to this table when the corresponding process
        in the sysApplElmtRun Table terminates.

        Entries remain in this table until they are aged out
        when either the number of entries in the table reaches
        a maximum as determined by sysApplElmtPastRunMaxRows,
        or when an entry has aged to exceed a time limit as
        set by sysApplElmtPastRunTblTimeLimit.  When aging out
        entries, the oldest entry, as determined by the value
        of sysApplElmtPastRunTimeEnded, will be removed first.

        The table is indexed by sysApplInstallPkgIndex
        (from the sysApplInstallPkgTable),
        sysApplElmtPastRunInvocID, and sysApplElmtPastRunIndex
        to make it easy to locate all previously executed
        processes of a particular invoked application that has
        been installed on the system."
    ::= { sysApplRun 4 }

sysApplElmtPastRunEntry OBJECT-TYPE
    SYNTAX        SysApplElmtPastRunEntry
    MAX-ACCESS    not-accessible
    STATUS        current
    DESCRIPTION
        "The logical row describing a process which was
        previously executed on this host as part of an
```

installed application. The entry is basically copied
from the sysApplElmtRunTable when the process
terminates. Hence, the entry's value for
sysApplElmtPastRunIndex is the same as its value
was for sysApplElmtRunIndex. Note carefully: only
those processes which could be associated with an
identified application are included in this table."
            INDEX   { sysApplInstallPkgIndex,
    sysApplElmtPastRunInvocID,
                    sysApplElmtPastRunIndex }
        ::= { sysApplElmtPastRunTable   1 }

    SysApplElmtPastRunEntry ::= SEQUENCE {
        sysApplElmtPastRunInvocID   Unsigned32,
        sysApplElmtPastRunIndex   Unsigned32,
        sysApplElmtPastRunInstallID   Unsigned32,
        sysApplElmtPastRunTimeStarted   DateAndTime,
        sysApplElmtPastRunTimeEnded   DateAndTime,
        sysApplElmtPastRunName   LongUtf8String,
        sysApplElmtPastRunParameters   Utf8String,
        sysApplElmtPastRunCPU   TimeTicks,
        sysApplElmtPastRunMemory   Unsigned32,
        sysApplElmtPastRunNumFiles   Unsigned32,
        sysApplElmtPastRunUser   Utf8String
    }

    sysApplElmtPastRunInvocID OBJECT-TYPE
        SYNTAX   Unsigned32 (1..'ffffffff'h)
        MAX-ACCESS not-accessible
        STATUS   current
        DESCRIPTION
            "Part of the index for this table, this value
            identifies the invocation of an application of which
            the process represented by this entry was a part.
            The value of this object is the same value as the
            sysApplRunIndex for the corresponding application
            invocation in the sysApplRunTable. If the invoked
            application as a whole has terminated, it will be the
            same as the sysApplPastRunIndex."
        ::= { sysApplElmtPastRunEntry 1 }

    sysApplElmtPastRunIndex OBJECT-TYPE
        SYNTAX   Unsigned32 (0..'ffffffff'h)
        MAX-ACCESS not-accessible
        STATUS   current
        DESCRIPTION
            "Part of the index for this table. An integer
            assigned by the agent equal to the corresponding
            sysApplElmtRunIndex which was removed from the
            sysApplElmtRunTable and moved to this table
            when the element terminated.

            Note: entries in this table are indexed by
            sysApplElmtPastRunInvocID, sysApplElmtPastRunIndex.

The possibility exists, though unlikely, of a
collision occurring by a new entry which was run
by the same invoked application (InvocID), and
was assigned the same process identification number
(ElmtRunIndex) as an element which was previously
run by the same invoked application.

Should this situation occur, the new entry replaces
the old entry.

See Section: 'Implementation Issues -
sysApplElmtPastRunTable Entry Collisions' for the
conditions that would have to occur in order for a
collision to occur."
::= { sysApplElmtPastRunEntry 2 }


```
sysApplElmtPastRunInstallID OBJECT-TYPE
    SYNTAX      Unsigned32 (1..'ffffffff'h)
    MAX-ACCESS  read-only
    STATUS      current
    DESCRIPTION
        "The index into the installed element table. The
        value of this object is the same value as the
        sysApplInstallElmtIndex for the application element
        of which this entry represents a previously executed
        process."
    ::= { sysApplElmtPastRunEntry 3 }

sysApplElmtPastRunTimeStarted OBJECT-TYPE
    SYNTAX      DateAndTime
    MAX-ACCESS  read-only
    STATUS      current
    DESCRIPTION
        "The time the process was started."
    ::= { sysApplElmtPastRunEntry 4 }

sysApplElmtPastRunTimeEnded OBJECT-TYPE
    SYNTAX      DateAndTime
    MAX-ACCESS  read-only
    STATUS      current
    DESCRIPTION
        "The time the process ended."
    ::= { sysApplElmtPastRunEntry 5 }

sysApplElmtPastRunName OBJECT-TYPE
    SYNTAX      LongUtf8String
    MAX-ACCESS  read-only
    STATUS      current
    DESCRIPTION
        "The full path and filename of the process.
        For example, '/opt/MYYpkg/bin/myyproc' would
        be returned for process 'myyproc' whose execution
        path was '/opt/MYYpkg/bin/myyproc'."
```

```
            ::= { sysApplElmtPastRunEntry 6 }


sysApplElmtPastRunParameters OBJECT-TYPE
     SYNTAX      Utf8String
     MAX-ACCESS  read-only
     STATUS      current
     DESCRIPTION
         "The starting parameters for the process."
     ::= { sysApplElmtPastRunEntry 7 }


sysApplElmtPastRunCPU OBJECT-TYPE
     SYNTAX      TimeTicks
     MAX-ACCESS  read-only
     STATUS      current
     DESCRIPTION
         "The last known number of centi-seconds of the total
         system's CPU resources consumed by this process.
         Note that on a multi-processor system, this value may
         increment by more than one centi-second in one
         centi-second of real (wall clock) time."
     ::= { sysApplElmtPastRunEntry 8 }


sysApplElmtPastRunMemory OBJECT-TYPE
     SYNTAX      Unsigned32 (0..'ffffffff'h)
     UNITS       "Kbytes"
     MAX-ACCESS  read-only
     STATUS      current
     DESCRIPTION
         "The last known total amount of real system memory
         measured in Kbytes allocated to this process before it
         terminated."
     ::= { sysApplElmtPastRunEntry 9 }


sysApplElmtPastRunNumFiles OBJECT-TYPE
     SYNTAX      Unsigned32 (0..'ffffffff'h)
     MAX-ACCESS  read-only
     STATUS      current
     DESCRIPTION
         "The last known number of files open by the
         process before it terminated.  Transport
         connections (sockets) should NOT be included in
         the calculation of this value."
     ::= { sysApplElmtPastRunEntry 10 }


sysApplElmtPastRunUser OBJECT-TYPE
     SYNTAX      Utf8String
     MAX-ACCESS  read-only
     STATUS      current
     DESCRIPTION
         "The process owner's login name (e.g. root)."
 ::= { sysApplElmtPastRunEntry 11 }
```

Of particular interest are those objects in the current and past element run tables that include the name and parameters associated with each process. We can use the name and parameter combinations to uniquely identify instances of processes that are really the same, except that they are doing work for different users. See, for example, *sysApplElmtPastRunName*. You may prefer to create names for these processes like "XYZ Company Premier," to reference the company for which the process is running and the service level being provided.

We could also use Definitions of Managed Objects for WWW Services [RFC 2594] to determine other important characteristics of Web server applications, such as the system services on which the Web server depends (e.g., the HTTP protocol).* Notice that *service* in the following table is defined more specifically that it has thus far. If we have a Web server that is important to a customer service, we can use this table, along with others, to understand at least some of the dependencies of the business service, such as the HTTP protocol:

```
-- The WWW Service Information Group
--
-- The WWW service information group contains information
-- about the WWW services known by the SNMP agent.

wwwService OBJECT IDENTIFIER ::= { wwwMIBObjects 1 }

wwwServiceTable OBJECT-TYPE
    SYNTAX      SEQUENCE OF WwwServiceEntry
    MAX-ACCESS  not-accessible
    STATUS      current
    DESCRIPTION
        "The table of the WWW services known by the SNMP
        agent."
    ::= { wwwService 1 }

wwwServiceEntry OBJECT-TYPE
    SYNTAX      WwwServiceEntry
    MAX-ACCESS  not-accessible
    STATUS      current
    DESCRIPTION
        "Details about a particular WWW service."
    INDEX       { wwwServiceIndex }
    ::= { wwwServiceTable 1 }

WwwServiceEntry ::= SEQUENCE {
    wwwServiceIndex             Unsigned32,
    wwwServiceDescription       Utf8String,
```

*Note that the MIB Module for Web services cited here is dependent on the Application Management MIB [RFC 2564].

```
        wwwServiceContact           Utf8String,
        wwwServiceProtocol          OBJECT IDENTIFIER,
        wwwServiceName              DisplayString,
        wwwServiceType              INTEGER,
        wwwServiceStartTime         DateAndTime,
        wwwServiceOperStatus        WwwOperStatus,
        wwwServiceLastChange        DateAndTime
}

wwwServiceIndex OBJECT-TYPE
    SYNTAX      Unsigned32 (1..4294967295)
    MAX-ACCESS  not-accessible
    STATUS      current
    DESCRIPTION
        "An integer used to uniquely identify a WWW service.
        The value must be the same as the corresponding value
        of the applSrvIndex defined in the Application
        Management MIB (APPLICATION-MIB) if the applSrvIndex
        object is available. It might be necessary to manually
        configure sub-agents in order to meet this
        requirement."
    ::= { wwwServiceEntry 1 }


wwwServiceDescription OBJECT-TYPE
    SYNTAX      Utf8String
    MAX-ACCESS  read-only
    STATUS      current
    DESCRIPTION
        "Textual description of the WWW service. This shall
        include at least the vendor and version number of the
        application realizing the WWW service. In a minimal
        case, this might be the Product Token (see RFC 2068)
        for the application."
    ::= { wwwServiceEntry 2 }


wwwServiceContact OBJECT-TYPE
    SYNTAX      Utf8String
    MAX-ACCESS  read-only
    STATUS      current
    DESCRIPTION
        "The textual identification of the contact person for
        this service, together with information on how to
        contact this person. For instance, this might be a
        string containing an email address, e.g.
        '<Webmaster@domain.name>'."
    ::= { wwwServiceEntry 3 }


wwwServiceProtocol OBJECT-TYPE
    SYNTAX      OBJECT IDENTIFIER
    MAX-ACCESS  read-only
    STATUS      current
    DESCRIPTION
        "An identification of the primary protocol in use by
        this service. For Internet applications, the IANA
```

```
            maintains a registry of the OIDs which correspond to
            well-known application protocols.  If the application
            protocol is not listed in the registry, an OID value
            of the form {applTCPProtoID port} or {applUDPProtoID
            port} are used for TCP-based and UDP-based protocols,
            respectively. In either case 'port' corresponds to the
            primary port number being used by the protocol."
        REFERENCE
            "The OID values applTCPProtoID and applUDPProtoID are
             defined in the NETWORK-SERVICES-MIB (RFC 2248)."
        ::= { wwwServiceEntry 4 }

wwwServiceName OBJECT-TYPE
        SYNTAX        DisplayString
        MAX-ACCESS    read-only
        STATUS        current
        DESCRIPTION
            "The fully qualified domain name by which this service
            is known. This object must contain the virtual host
            name if the service is realized for a virtual host."
        ::= { wwwServiceEntry 5 }

wwwServiceType OBJECT-TYPE
        SYNTAX        INTEGER {
                        wwwOther(1),
                        wwwServer(2),
                        wwwClient(3),
                        wwwProxy(4),
                        wwwCachingProxy(5)
                      }
        MAX-ACCESS    read-only
        STATUS        current
        DESCRIPTION
            "The application type using or realizing this WWW
            service."
        ::= { wwwServiceEntry 6 }

wwwServiceStartTime OBJECT-TYPE
        SYNTAX        DateAndTime
        MAX-ACCESS    read-only
        STATUS        current
        DESCRIPTION
            "The date and time when this WWW service was last
            started. The value SHALL be '0000000000000000'H if the
            last start time of this WWW service is not known."
        ::= { wwwServiceEntry 7 }

wwwServiceOperStatus OBJECT-TYPE
        SYNTAX        WwwOperStatus
        MAX-ACCESS    read-only
        STATUS        current
        DESCRIPTION
            "Indicates the operational status of the WWW service."
        ::= { wwwServiceEntry 8 }
```

```
wwwServiceLastChange OBJECT-TYPE
     SYNTAX        DateAndTime
     MAX-ACCESS    read-only
     STATUS        current
     DESCRIPTION
        "The date and time when this WWW service entered its
         current operational state. The value SHALL be
'0000000000000000'H if
         the time of the last state change is not known."
::= { wwwServiceEntry 9 }
```

These tables suggest potential policies that look for specific instances in all tables, not just to determine if the desired services are running, but also to discover current state. In the event of a "missing" application, the system could use past run tables to find out when the application exited.

The system might also send a TRAP or INFORM to the service management application when processes are missing, or when they've stopped and restarted a certain number of times.

Note that the term *elements* in [RFC2287] denotes a process. This is a narrower definition than that used in an SNMPCONF context, but is not terribly problematic, because 2287 elements are also likely to be SNMPCONF elements.

Part of the power of policy-based service management is that it gives us a view of services as a collection of elements: We do not have to poll each and every process in a system to find out if the elements of interest are performing correctly. Because we know exactly which elements (instances) are of interest, we can retrieve only those we care about. To take this one step farther, we might even avoid polling the interesting elements by causing a notification message to be sent when attention is warranted. Not only does this dramatically reduce the amount of data moving through the network, it greatly reduces the amount of data service management applications network operators must process. It also means that fewer extraneous events posted to the operational staff, thus freeing them up for high-priority problems instead of sending them off to track down "non-problems."

Step 3: Service-level Reporting

What is the distinction between monitoring and reporting? In the previous section, we looked at information that monitored the relative health of the service (were all the processes running that should be?). For service-level reporting, we want to collect data to help quantify the amount of the service provided. In our simple example, we could count the number of pages the Web server provides over some period of time. We could then report the number of pages served to accounting and billing systems to trigger the generation of correct bills or credits. A number of tables in [RFC2594] can assist service-level reporting for our Web server:

```
wwwDocBucketTable OBJECT-TYPE
    SYNTAX        SEQUENCE OF WwwDocBucketEntry
    MAX-ACCESS    not-accessible
    STATUS        current
    DESCRIPTION
        "This table provides administrative summary
        information for the buckets maintained per WWW
        service."
    ::= { wwwDocumentStatistics 3 }

wwwDocBucketEntry OBJECT-TYPE
    SYNTAX        WwwDocBucketEntry
    MAX-ACCESS    not-accessible
    STATUS        current
    DESCRIPTION
        "An entry which describes the parameters associated
        with a particular bucket."
    INDEX         { wwwServiceIndex, wwwDocBucketIndex }
    ::= { wwwDocBucketTable 1 }

WwwDocBucketEntry ::= SEQUENCE {
    wwwDocBucketIndex            Unsigned32,
    wwwDocBucketTimeStamp        DateAndTime,
    wwwDocBucketAccesses         Unsigned32,
    wwwDocBucketDocuments        Unsigned32,
    wwwDocBucketBytes            Unsigned32
}

wwwDocBucketIndex OBJECT-TYPE
    SYNTAX        Unsigned32 (1..4294967295)
    MAX-ACCESS    not-accessible
    STATUS        current
    DESCRIPTION
        "An arbitrary monotonically increasing integer number
        used for indexing the wwwDocBucketTable. The index
        number wraps to 1 whenever the maximum value is
        reached."
    ::= { wwwDocBucketEntry 1 }

wwwDocBucketTimeStamp OBJECT-TYPE
    SYNTAX        DateAndTime
    MAX-ACCESS    read-only
    STATUS        current
    DESCRIPTION
        "The date and time when the bucket was made
```

```
                           available."
                  ::= { wwwDocBucketEntry 2 }

        wwwDocBucketAccesses OBJECT-TYPE
            SYNTAX       Unsigned32
            MAX-ACCESS   read-only
            STATUS       current
            DESCRIPTION
                "The total number of access attempts for any document
                 provided by this WWW service during the time interval
                 over which this bucket was created."
            ::= { wwwDocBucketEntry 3 }

        wwwDocBucketDocuments OBJECT-TYPE
            SYNTAX       Unsigned32
            MAX-ACCESS   read-only
            STATUS       current
            DESCRIPTION
                "The total number of different documents for which
                 access was attempted this this WWW service during the
                 time interval over which this bucket was created."
            ::= { wwwDocBucketEntry 4 }

        wwwDocBucketBytes OBJECT-TYPE
            SYNTAX       Unsigned32
            MAX-ACCESS   read-only
            STATUS       current
            DESCRIPTION
                "The total number of content bytes which were
                 transferred from this WWW service during the time
                 interval over which this bucket was created."
            ::= { wwwDocBucketEntry 5 }
```

Device-Level Service Monitoring and Reporting

Throughout this chapter, we've employed an example focused exclusively on a single device. Using SNMP service-level management, all that can be reliably determined from inside this single box is its internal state and utilization information. In a fully deployed service management system, there are many application or Web server systems, and each must be monitored along with other devices critical to the delivery of the service. The service management system is responsible for correlating information collected from all the managed devices to determine if a service had, in fact, been disrupted at any point. Yet, the fact that service management software must perform this aggregating function in no way reduces the importance of the simple example. Without policy-based service verification and reporting, management software would be far more complex and have to collect far more data far more frequently, thus unnecessarily increasing demand on the managed device, the network, and the service management software itself.

PART

5

Applying Design Principles

Part 5 focuses on the specific recommendations for the implementation of software for managed devices and management applications. The benefit of adding specialized SNMP subagents is explored is Chapter 10 and illustrates the value of distributing intelligence from management systems to managed devices.

In Chapter 11, we get to the punch line; an architecture for a complete service management system that can function across vendors, technologies, and access methods.

In Chapter 12, we see a complete system that includes both management applications and managed devices with the management infrastructure I have advocated. With this we can see how all the parts of the management system function together to produce a complete service management environment.

No single piece of software, however well conceived, implemented, and deployed, is likely to meet all requirements for all customers. For this reason, a complete management system should have an effective set of external interfaces that humans and other applications can use. These issues are described in Chapter 13.

Finally, we conclude in Chapter 14 with a look at the future and the various roles that vendors, users, and standards bodies can play in shaping that future.

CHAPTER 10

Specialized Function Subagents

Because SNMPCONF technology operates on vendor-proprietary and standard MIB Objects, it can be used with any object, extant or future. It represents one possible way that we can improve the quality of management in an SNMP-based infrastructure. Fortunately, not all improvements are dependent on the Policy MIB Module developed in the SNMP-CONF Working Group and the software written based on that document. Many other techniques are available for the creation of SNMP agents to help us deliver SNMP-based service management systems. Now we'll turn to some of these other techniques, which can improve efficiency and make it easier for operational personnel to support business goals. In the next sections, we discuss MIB Objects (and the SNMP agents that realize them) that can provide added value in the areas of fault and event/alarm generation; configuration verification, and testing; performance and accounting, through statistics aggregation; and service-level verification through distributed analytic functions.

Fault and Event/Alarm Generation

Before I describe specific techniques for improving alarm and event management, I'll clarify the two terms, as we use them here:

* An *event* is any condition that management systems have detected in a system. A common example of an event is a port failure, such as described in the section, "Dealing with Notifications," in Chapter 8.
* An *alarm* is the enunciation of one or more events. "Enunciation" could be as simple as writing to a log, putting up a message on a screen, or having the system dial a pager. Sometimes there is a one-to-one relationship between an event and an alarm, whereas in other cases many events of the same kind must be generated (generally within some time interval) before an alarm is raised. In still other cases, events of different types must first be observed by the alarm/event management software before the system sends an alarm. Note that these rules may be built in or customizable by the user. Good software in managed devices provides a bit of flexibility in this area. The greatest amount of flexibility in defining an alarm is most often found in fault management applications rather than in the managed devices, which frequently just send notice of events.

I've somewhat simplified these definitions, but they will serve this discussion well. The International Telecommunications Union [ITU] (see the useful URLs list in Appendix C) has devoted a great deal of research and study to this topic, which is quite complex. A search for *event* on the ITU Web site returned 85 different references! In his book *Telecommunications Network Management*, Haojin Wang [WANG] has several excellent discussions on event and alarm management in a TMN environment, and the general principles in his book can be applied broadly to problems in the IP management domain. The IETF has only recently put a good deal of study into this important area. Some work is taking place in the Distributed Management Working Group [DISMAN], which is most likely to adopt terms and definitions similar to those of the ITU.

A perennial problem for operational staffs is that they have too many, as opposed to too few, SNMP TRAP or INFORM-based* alarms to handle. In this case, there is a one-to-one relationship between an event and an alarm. Each event—let's say it's an interface failure—results in the transmission of an SNMP TRAP or INFORM PDU that the management software raises as an alarm. It most often raises the alarm on the graphical management application by turning the color of a device red, beeping, and adding an entry to a scrolling log window or some combination of all three. If you've ever wondered why there is such a variety of alarm correlation software on the market, remember that its primary function is to take events in the form of SNMP TRAP or INFORM PDUs, screen out duplicates, and correlate information to reduce the number of alarms displayed on operator consoles. Remember the design discussion on management software for managed devices and some of the issues regarding asynchronous notifications: much of this alarm-suppression software exists because the managed devices send out too many duplicate or low-value messages. In this chapter, we discuss some techniques to address this problem.

The event/alarm correlation software to deal with duplicate events, as well other important features, can offer significant benefits, but it does have a price tag. Costs associated with the acquisition and deployment can be reduced by using the more intelligent agent implementations suggested earlier. Such implementations could slash the amount of time

---

*In an earlier note we talked about SNMP versions and how different PDUs and the SMI evolved. Rather than clutter up the text with SNMPv2-TRAP-PDU, and so on for each version or InformRequest-PDU, I will often use TRAP for all versions of TRAP that can be generated and INFORM as a shorthand for InformRequest-PDU. I spell things out where there is a particular point to be made about a protocol interaction.

it takes to identify and repair failures, and the costs associated with them. Let's examine some beneficial strategies you can employ when developing an SNMP agent that sends asynchronous messages (TRAP or INFORM PDUs).

Grouping Events Based on a Hierarchy

In many cases, simply grouping events based on hardware or software hierarchy can reduce the number of events detected by management software and the resulting display of alarms for operators. In fact, many network devices exhibit a natural hierarchy of hardware and software. For example, many systems have slots that accept cards. Larger systems get more complex with the cards aggregating into chassis, and then into the 7-foot racks common in telecommunications facilities.

Recall the discussion about too many NOTIFICATION messages in Chapter 8. This discussion provides at least one technique for avoiding sending out too many messages as a result of a problem with a four-port card. Figure 10.1 shows a simple system with several cards. In this case, network interface cards (NICs) can be inserted into the chassis. Each NIC in our example has four ports. Notice that one of the ports, port #4 on card #2, has been shaded to indicate a failure. In this case, an appropriate TRAP or INFORM can be sent to carry important information to the network operator. This is the easy case.

Now notice that all the ports on card #1 have been shaded, and so has the background of the card itself. In this case, the card probably failed, as opposed to the unlikely event that all four ports failed at exactly the same time. If the management software *inside this device* has enough intelligence, it can send a single message to inform the management system of the card failure, and, by extension, the failure of all four ports. This is not to say there can't be edge conditions where it is preferable to send multiple messages, one for each port, than to indicate the board is down. For the most part, though, a single message reporting board failure communicates the problem more accurately and requires less management software to figure out what is wrong. The result is a more effective NOC.

Figure 10.1
Event aggregation.

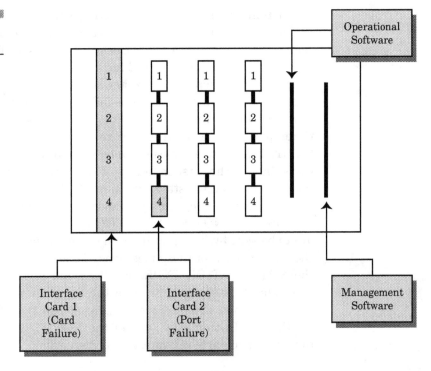

Centralized Control of TRAP and INFORM PDUs

No matter what level of aggregation hierarchy-based software provides, there still could be too much data flowing from a managed device. One way to control it is to create a centralized "agent" to manage TRAP and INFORM PDU generation. Suppose we have a Web server containing management software configured to send a notification message when the CPU utilization of certain processes reaches a predetermined level. Several types of control can be helpful:

※ **Number per type of notifications per unit of time**—Event/alarm information is defined in the Structure of Management Information Version 2 (SMIv2) [RFC2578] using the NOTIFICATION-TYPE MACRO. An effective method of reducing unwanted generation is to assign a portion of the SNMP infrastructure inside the managed device to track NOTIFICATION-TYPE information sent out in either TRAP or INFORM PDUs. Say that process utilization in our Web server remains elevated over the trigger level for an extended period of time. Depending on how the NOTIFICATION-TYPE is defined, this condition could cause a lot of messages to be sent from the server to the management systems that have been configured to receive them. A similar condition would exist where a TRAP or INFORM PDU is sent every time one of the ports fails. In the case of an intermittent problem, where the interface fails and comes back, perhaps many times per minute, we'd expect to generate many such messages. Clearly there can be times when one would want to receive all transitions, but in the vast majority of cases it's sufficient to know the interface is changing states at a fairly high rate.

Software on the managed device can control the rate at which instances of a NOTIFICATION-TYPE (such as our interface failure) are to be sent by fixing an interval (once per minute, for example) for the transmission of these messages. Sometimes the rate is defined in the NOTIFICATION-TYPE definition, but it often is not, and software should step in to assume the function to limit unnecessary transmissions. Earlier, we presented the Interfaces Table from [RFC2863]. In that RFC, one of the NOTIFICATION-TYPE definitions is particularly relevant to this discussion:

```
linkDown NOTIFICATION-TYPE
     OBJECTS { ifIndex, ifAdminStatus, ifOperStatus }
     STATUS  current
     DESCRIPTION
             "A linkDown trap signifies that the SNMP entity,
             acting in an agent role, has detected that the
             ifOperStatus object for one of its communication
             links is about to enter the down state from some
             other state (but not from the notPresent state).
             This other state is indicated by the included value
             of ifOperStatus."
     ::= { snmpTraps 3 }
```

Notice that there's no limit on the number of times a message can be sent from a device with an instance of the NOTIFICATION-TYPE. For this class of object, software that controls the rate of message generation over some period of time can be useful. Ideally, users could adjust this in the field by providing a value to a MIB object supplied by the vendor for this purpose.

■ **Number per destination per unit of time**—A variation or refinement on the previous point is to control the rate of transmission by specific destination. Not all systems have the same requirements for the consumption of failure and other type of event information. This is particularly true in a distributed management environment, where some systems may emphasize the fault management function and others performance or service-level management. The fault management system might want to receive notification data at a higher-than-typical rate to scan for possible sources of the difficulty. On the other hand, a service level system doesn't need to know more than that an interface failure has occurred or has persisted for a specified period of time. The level of granularity required here is going to be determined by how the service level agreement is defined.

- **Absolute number of all types generated per unit of time**—A third type of aggregation limits the total number of messages of all types (and destinations) a device can send. This type is particularly valuable when the device must send the messages over a slow link, as it might when a branch device has been configured to send messages to a centralized management system at a remote network operations center. Supposing the system is still able to function, we do not want to absorb more bandwidth than necessary of the vital link that connects the branch office to the network core. Putting a ceiling on the number of messages per unit of time conserves bandwidth.

  Note that sophisticated event and alarm correlation software is helpless to deal with the problem of too many notifications being sent over a slow link—they can filter them out from the operator's perspective only. By the time the messages arrive at the event and alarm correlation software, it is already too late—the valuable bandwidth has been expended.

- **Absolute number of retransmissions per unit of time**—When a TRAP message is sent, no response is due from the intended recipient, so the sending device has no way of knowing if the message was received. In this case, limits on the number of retransmissions of a specific TRAP message for a specific period of time can help. When an entity receives an InformRequest-PDU, correct behavior is to reply with a Response-PDU [RFC1905]. The Response-PDU has proved beneficial and can be used to avoid some unnecessary retransmission, even so it makes sense to consider additional limits. The system that generates the INFORM still can't determine the message's fate without a response: it may have been lost on the way to the management system, or it may have been lost in the management system, or the Response-PDU may have been lost on transmission back to the managed device. For this reason, the system should contain configuration options to restrict retransmissions to managers from which Response-PDUs have not been received for some number of outstanding INFORM messages.

Event Correlation inside Managed Devices

The most sophisticated type of correlation is when a managed device itself examines several event types and, based on that analysis, decides what notification information to send and how frequently to send it.

The system in Figure 10.2 might have several SNMP subagents to instrument different portions of the system. One subagent may implement System-Level Managed Objects for Applications (indicated by the "Process Events" blocks). It might also handle objects from the Application Management MIB. Another subagent may implement Definitions of Managed Objects for WWW Services (indicated by the "Web Events" blocks). Interface events from the Interfaces Group MIB are most likely to be implemented along with other common MIB Objects by agent software within the base system.

Figure 10.2
Event correlation.

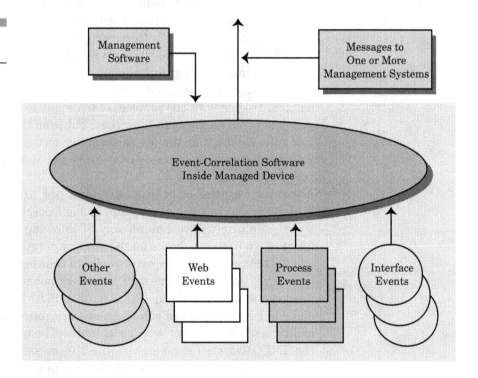

If a device interface fails or starts to fluctuate between states in which it can effectively send and receive packets, the system will be unable to serve Web pages. Similarly, if one of the main Web processes fails, Web services will be degraded or halted. If the machine is in service as a virtual Web server, where many Web servers are running more or less independently but share a common network interface, an interface failure can cause the generation of many apparently unrelated events—unless you know about the interface failure, that is.

Let's create software that could be configured to correlate these events—not unlike the function many root-cause analysis products perform today. The important distinction is that the managed device does the correlating, thus easing the load on the external management system infrastructure.

Some of you may have never seen an IP-based network device with this level of event correlation capability, because management software with this functionality is fairly recent. Processing requirements for these functions are not insignificant, and for some managed systems such software may be too costly. But don't be quick to decide that's true for you. It's a good investment for the majority of network devices such as medium- to large-scale routers, workstations, servers, and the like. If we are to develop and deploy high-quality and high-value services, management software functions on devices, as well as service management applications will have to be upgraded.

Configuration Verification and Testing

Most new management technology targets configuration management. For these technologies, SNMCONF included, some attention is also paid to basic error checking and the reporting of invalid data. If your goal is to increase the likelihood of a successful configuration change across a large number of devices, however, "some attention" is insufficient. This is even more problematic when the devices span a range of technologies and vendors.

No matter how complete the configuration management applications are or how well tested, it is a practical impossibility for them to employ all the semantic checking provided in each network device's operational code. For example, many routing systems parse configuration information and, using a fairly complex set of algorithms, determine if it is correct. If it is not, they must also determine which statements in the configuration file are in error. The only way for management software to achieve the same level of accuracy is to implement most of the same code as the operational software. Clearly, this is not a viable solution, especially for software developers who support a range of vendors. Yet configuration management applications must somehow be equipped to ensure that messages exhibit correct syntax for each managed device, and correct semantics to the greatest degree possible.

The management software on managed devices can apply several techniques to real benefit if we want to verify before putting the new configuration information online:

- **Take the subsystem out of service**—Using this approach, the new configuration is installed on the device and the operational code is temporarily taken offline while it's being loaded and tested with new configuration information. A good example of an SNMP MIB Object that performs this function is found in the OSPF Version 2 Management Information Base [RFC1850].

```
ospfAdminStat OBJECT-TYPE
    SYNTAX      Status
    MAX-ACCESS    read-write
    STATUS      current
    DESCRIPTION
       "The  administrative  status  of  OSPF  in  the
       router.   The  value 'enabled' denotes that the
       OSPF Process is active on at least  one  inter-
       face;  'disabled'  disables  it  on  all  inter-
       faces."
::= { ospfGeneralGroup 2 }
```

The effectiveness of this approach depends on how the agent and operating software are implemented. Note that this object does not specify what happens when the object is set to "disabled." If you were building a system to support the function of background configuration checking, it would be advantageous to develop additional MIB Objects to further refine offline behavior.

- **Create another instance of the software and feed it the new configuration information, letting the operational code verify its syntax and semantics while the original configuration is still running**—Taking a process offline is a fairly drastic measure and may not be appropriate in a lot of cases. In environments where it is technically feasible, invoking another instance of the service or subsystem being reconfigured may be preferable. Using this approach, the operational code reads the new configuration information and performs its normal verification process. Errors are logged and reported to the management system for corrective action before going live on the device.

Performance and Accounting, Statistics Aggregation

The requirements of service level, accounting, performance, and capacity planning reporting can often be met using aggregated information. For this purpose, aggregation of data can be either temporal or a combination of multiple instances.

* **Temporal aggregation of statistics**—If you have a MIB Object that is defined as a counter on a high-speed interface, normal practice is to poll the values of each interesting instance. This retrieval is often repeated many times an hour, every hour of the day. Originally this approach was motivated by the fact that some objects using 32-bit counters would wrap in a short period of time; rapid polling was necessary to avoid missing a wrap of the counter and the valuable information it provides. With the advent of 64-bit counters in SMIv2 [RFC2578], the need for rapid polling is somewhat reduced. However, there are still many objects that might benefit from the localized aggregation of their values, particularly because it places only a small incremental load on the managed device and can significantly reduce processing load on management applications. The processing and consolidation of polled data in a management application is one of the most hardware resource–consuming tasks in these systems. For example, if we were to create an object that represented the number of IfInOctets seen on an interface over some user-configurable 24-hour period, we could significantly reduce the number of requests for the regular ifInOctets object. Of course, if the application needs fine-grained detail, this will not work.
* **Combination of multiple instances**—As systems get larger, the number of instances increases. Aggregating multiple instances of a specific object type such as ifInOctets into a single aggregate object can condense the amount of data to move between a single managed device and a manger. A router with sixteen interfaces, for instance, is likely to have sixteen instances of ifInOctets. Many management systems poll for these objects several times per hour. If just half of these

interfaces don't have to report with a fine level of detail (e.g., they are not supporting customers with service level agreements) and we poll twice a day, traffic and load on the management system can be significantly improved. Note that counter wrap is not a problem here because the aggregation subagent polls at whatever frequency is required to ensure it does not miss data.

To understand just how much we can reduce polling and subsequent processing with this technique, compare it to the number of polled instances of just a single object type such as *ifInOctets*. In the example above, we had a router with sixteen interfaces, which is relatively small in many environments, and we elected to collect data on eight of them. If you normally poll every 15 minutes, that's four times per hour and 96 times per day for each interface, or 768 instances of a value for *ifInOctets* data in a 24-hour period. If you aggregate the data for each interface during that 24-hour period, however, you'll collect eight data objects. Each one of these eight data objects represents, in this case, the total number of *ifInOctets* seen on one of the eight interfaces we have selected for aggregation over a 24-hour period. This is temporal aggregation. Suppose we combine these eight interfaces into one aggregate and poll once per day. We now have one data element compared to 768. This is an example of multiple instance combination—in this case, the combination of the eight interface instances. Because it is hard to define all possible objects that might be useful in our current discussion, configuration objects are needed in the managed device to let operators specify what instances are to be combined and over what periods of time, if any.

This approach may not be practical in all cases, but it can be of significant value in some circumstances. Many reasons exist for why you would want more fine-grained interface counter statistics than one for all the interfaces for each 24 hours. This rather extreme example is used simply to show what is possible.

A specialized subagent, based on standard or vendor-specific objects already defined, can be created to realize at least some of these aggregated objects. Our special-purpose subagent then populates these aggregated objects. Ideally, it can be configured to control the aggregation intervals and the instances to be aggregated. An additional benefit of this approach is that the fine-grained data we've bypassed is still on the system, in subagents that realize the basic objects. When it's needed to help isolate and resolve faults, it can be polled as often as needed for the problem resolution process.

A subtle point that this discussion emphasizes is the importance of an integrated management application. If we have separate applications for fault, data collection for general capacity planning (say, for each router and each interface), and service-level agreements, we almost certainly do too much polling of the managed devices, even if they have good aggregation facilities. One reason for this is that without knowing how elements are configured and what services and customers use them, the polling software must fall back to its traditional approach; that is, to collect everything from everywhere, all the time. For scalable service management environments we need integrated management applications that work with "smart" managed devices having some of the facilities we describe here.

Service Level Verification, Distributed Analytic Functions

If aggregation distributes some of the work of data collection to managed devices, other techniques can distribute some of the work for service-level verification. As we've observed previously, it's easier to establish that service levels are being met if you have the configuration information associated with the counter and state information. We could send:

- **Service-level notification messages based on state**—For a service that depends on several components all being in an up state, we might want to cause a notification message to be sent when one of the mandatory components fails. In fact, notification messages could differ according to the severity or importance of the failure. If a backup interface fails, that is important, but not as important as an interface currently carrying traffic for which there is no backup. For this, you'll need a special agent that implements NOTIFICATION-TYPEs. It could be configured manually or, in a system that has a Policy Agent [PM], it could be configured dynamically based on the policies installed.

- **Service-level notification messages based on utilization**—Accurate billing depends at least in part on having information about when a particular billable entity exceeds committed rates. If a polling-based approach is used for this purpose, a great deal of additional load falls on managed devices, the network, and management applications for basic accounting. To record utilization, we do not need very fine-grained detail. Assuming billing based on the amount of data forwarded on behalf of some billable entity, all we need is to reliably record information at some reasonable interval, perhaps each hour. We then retrieve the totals daily. Sometimes service-level agreements will be written so that, in addition to total utilization over a period of an hour or day, they will also have a restriction on a peak rate of transfer per some unit of time. Imagine we have an agreement that promises a certain level of performance and guaranteed data transfer over the course of an hour. Clearly, the amount of resources necessary to transfer that amount of data will be far greater if all the data is moved during a single 5-minute interval as opposed to more gradually over an hour. Bursts like this happen all the time. The question is what to do about them. In some cases, the customer may be willing to pay an extra charge if the service provider can move larger amounts of data over a shorter period of time than normally expected. As it turns out, at the end of an hour, more, less, or about the same amount of data might have been transferred when we have a peak. Most times, more data will have been moved. Naturally, the service provider wants to get paid for providing the capability for supporting these bursts—they could be an important source of revenue. To do this, you must track these conditions.

If we want to track the peaks and charge extra for them once the peak rate promised in the SLA has been exceeded, we must make provision in our service management software. In Figure 10.3, Peak versus Utilization Data Collection, a customer has a service agreement that permits peaks above 15 mb for a maximum of one 5-minute interval at a time. The service-level agreement calls for the customer to pay a surcharge for each interval exceeding 15mb of utilization that follows another such interval. Utilization data could presumably be gleaned from this peak data, but that is a costly way to collect data. A smarter way is to define a NOTIFICATION-TYPE to be sent whenever limits have been exceeded. Again, this should be configurable on the device. The receipt of these messages can be logged and fed into the accounting system, which will help generate the correct bills.

■ **Service-level warnings (latency, etc.)**—Smart agents are another type of service level notification that can save a great deal of data collection and analysis work at the management system level. These agents monitor the qualitative aspects of a service and tell the service management application when the qualitative metrics of the service level agreement are not met, or have not been met for a configurable period of time. Here again, the real-time nature of the data can be

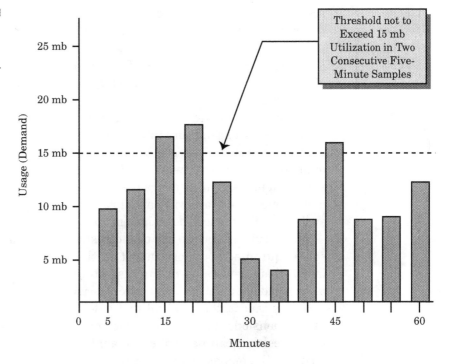

Figure 10.3
Peak versus utilization data collection.

significantly improved while reducing the load on the network management system.

Designing for Aggregation

Each of the approaches we have discussed has an associated implementation cost per managed device. Additionally, you will incur ongoing CPU and memory costs when enabling these functions in a managed device. The service management application must be written to take advantage of these features (most systems are not deployed with them). As service management becomes a more important consideration, of course, it is likely that these features will also become more important. It is up to the system designer of the management software infrastructure in each managed device to determine the correct balance. Fortunately, the IETF Distributed Management Working Group has some work that can help in this area. Three documents in particular might be useful building blocks to the functions we have just discussed. These are Definitions of Managed Objects for Remote Ping, Traceroute, and Lookup Operations [RFC2925], Event MIB [RFC2981], and Notification Log MIB [RFC3014]. The IETF is are also dealing with events and alarms; this work is worth tracking on the DISMAN Web page (see Appendix B).

There is little point investing in managed devices software without a corresponding expenditure in management applications that can use the data it generates. This is not to suggest that equipment vendors must implement all this software. To the contrary: if the principles of good service management software development are applied, investments in management software will pay off richly because more functions will be available with reduced development cost.

That said, we have a bit of a chicken-and-egg problem going: until some equipment vendors invest in the management infrastructure inside their devices, there's no way to justify the development of the management application software described in the next chapter. Even with this caveat, the material that follows can be of value in two important dimensions. First, for management software developers, there are principles that can be applied broadly even in the absence of improvements to the management infrastructure of network devices. Second, for management software purchasers who plan to deploy it in their networks, understanding the design principles can help them tackle scale issues and deploy more effectively.

CHAPTER 11

Management Applications— Architecture and Basic Components

Any implementation of an effective service management system must contain all of the components shown in Figure 11.1. Although the details may vary from implementation to implementation, the basic requirements will not.

Figure 11.1
Service management
software functional
components.

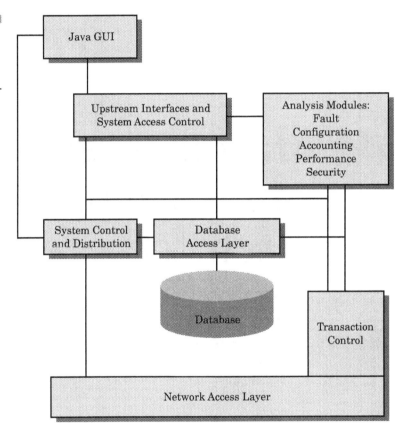

A detailed review of external interfaces, including the graphic user interface (GUI), is covered in the next chapter, which builds on the other components in the diagram and has a special set of issues. The block representing upstream interfaces and system access control provides critical security features for the entire service management system, controlling not only who has access, but also which managed devices can be accessed. These security issues are discussed in a later section.

Network Access Layer

The network access layer is responsible for all communications of any type with all the devices under the aegis of the management software. This layer is not responsible for communication initiation with other management systems, although some of the mechanisms of this layer (such as SNMP or FTP) may be used. The capability for communication initiation with other management systems exists in the upstream interfaces and the system access control layer.

The network access layer does not include those features resident in the operating system that enable low-level communications. Low-level communications are directly supplied by the operating environment. So, while software with SNMP capabilities is part of the network access layer in a management system, the low-level UDP code that SNMP uses is not.

If a management system uses telnet or SSH® [SSH] to communicate with CLIs on managed devices, the software is loosely considered part of the network access layer as well, and can use operating system services as needed. A distinction between telnet and SNMP is that the former is often part of the operating environment, whereas software that performs SNMP manager functions generally is not. True, most modern systems provide SNMP agent support, but very few provide software to perform SNMP GET and related operations, which are historically thought of as management functions.* Most large-scale management software projects begin with the acquisition of a commercial SNMP software suite that can perform all the SNMP functions required of a manager, such as SET operations and SNMP GetBulk [RFC1905]. This software also has high-quality technical support to help the development team with scale and other issues.

Support for Multiple Access Protocols

The primary function of the network access layer is to support the individual protocol details of each access method. In the case of SNMP, this

---

*Many open systems based on one UNIX variant or another now ship very basic SNMP manager-type functions that include GET, SET, etc. The point remains that this "free" software is not capable enough in many cases to act as a foundation for the network access layer for a service management system. It can often be helpful to very knowledgeable users in very small environments. This free software varies with regard to features built in for scale, reliability, and technical support.

layer is responsible for generating the appropriate SET or GET request (GetBulk, Get, etc.) [RFC1905] and receiving and processing notifications as they come in. Protocol-level controls can be used to set time out and retry parameters, adjusting this layer to fit the particular characteristics of a network environment.

The network access layer receives requests from the transaction control system and distributes them to the correct access method for each managed device. Some managed systems support only a CLI-based configuration approach, whereas others (e.g., modern cable modem systems) are primarily configured using SNMP. In almost all cases, however, usage and other statistics are generally available via SNMP. That's also true for hybrid systems, where configuration activities are split between the CLI and SNMP, as would be the case if a CLI-based device had an SNMP parameter to initiate a file transfer of new configuration information.

Many management systems are written with only a single management protocol in mind, which may work for specific models from a specific vendor, but is problematic if you want your management system to manage multiple types of equipment with multiple management access methods. Even if your systems all use the same access method, there will be significant variations to handle. To address this problem, the entire management system should be designed to support the multiple access methods found in this layer. Although not in common usage, the idea of using [XML] to transfer configuration information is possible. XML-RPC has been proposed as a way for communicating the XML-based configuration data to managed devices. If this approach catches on, the network access layer is where this method of communication would be integrated into the system. This illustrates the importance of flexibility at this layer; it also illustrates the importance of a flexible object model with regard to access methods. (Figure 11.3, showing Object Relationships later in this chapter, shows a system with CLI, File, and SNMP support. It would be relatively easy to add XML to such a system.)

Transaction Control

In discussing transactional integrity in Chapter 8, Figure 8.8 illustrated the levels of transaction in a service management system. The network access layer in the management software is responsible for only two of those levels, as they relate to individual devices:

▓ **Protocol-level transactions, such as those used by SNMP**—If we were to generate an SNMP PDU to collect management information from a network device, the network access layer would be responsible for deciding which PDU to use when there is a choice to be made. It would also be responsible for handling failures and performing retries and time-outs for each PDU. Appropriately, in the network access layer the security specified for each management protocol is applied. In the case of unrecoverable errors, the network access layer must be able to pass the error information up to the transaction control level, so that managers can determine whether a rollback to the previous state is appropriate on a device where protocol failure occurred. In some cases, it might also be appropriate to roll back transactions on all the devices that were part of the larger transaction that included the failed device. That decision would be made above the network access layer.

If you're evaluating software from an operational perspective, there are two important considerations at the network access layer. Find out whether this portion of the system is equipped with configuration controls, so that it can be adjusted for each device and the network as a whole. Then get a good idea of how well this layer is integrated with the rest of the management system. If it doesn't reliably pass failures up to higher layers, the system as a whole may not operate effectively. Some vendors offer error logs, which can be helpful, but don't do the whole job. The *entire system must be connected*, so that information about a failure at one level is passed to other parts of the system that need to know. This detail is too often overlooked.

▓ **Completing operations at the device level**—The second level of transaction handled by the network layer is performing operations for an entire system. To configure or retrieve data from a managed device, the network access layer may have many individual protocol-level actions to complete. So, although this level does not need a network-wide perspective on a transaction, it should know when it has completed all of the transactions requested for a given device within the context of a larger transaction. There may be multiple large-scale transactions in effect at any moment, and more than one might relate to the same device. This arrangement may strike you as messy, but using multiple transactions per managed device in the context of some larger network transaction, such as adding capacity for a customer who already has a service, yields better error recovery and control.

Transaction Control Layer

The transaction control layer is especially important for configuration operations. It is responsible for accepting requests from the configuration analysis modules and passing them on to the network access layer. It also passes the results of operations at the network access layer back to the configuration analysis layer and (potentially concurrently) to the database access layer. The functions included in the transaction control layer are:

* **Tracking each network-wide transaction**, such as reconfiguring many network devices at once or collecting accounting information at a predetermined time. The transaction control layer does not maintain lists of what is to be reconfigured or from which elements to collect data at specific intervals. That information is in the database and the various analysis modules and is given to the transaction control layer as needed. The transaction control layer simply receives the transaction requests and maintains the state of all ongoing transactions being serviced by the relatively stateless network access layer.
* **Transaction status reporting**—Some transactions may take a long time, even hours. The rest of the system (including the users at the graphic interface) will need updates and status information. This layer must be prepared to respond to requests for status information and, in some cases, to provide updates at configured intervals or when certain conditions (such as failures) must be reported.
* **Transaction completion reporting**—When a transaction has been completed, the results must be passed to the requesting analysis module and frequently to the database. A transaction could be for a single device or several subsystems on a single device. Most often it includes many devices at one time.
* **Transaction error/exception reporting**—When failures or errors occur, they should be logged with real-time messages, passed on to the requesting analytic module.
* **Transaction logging for rollback**—It is not the responsibility of the transaction control layer to make decisions about rollback. These should be handled by higher-level software. However, it is this layer's responsibility to keep track of all operations within a transaction and their state. When a higher-level program decides a rollback should be attempted on some or all devices, the information will be available. One reasonable place to keep some or all of the transaction log information is in temporary tables in a database, which makes it easier to update the database when the transaction is complete.

Many configuration systems have some of these transaction control layer functions. Few, if any, have them for multiple types of systems and vendors at one time. This is a particularly challenging aspect of developing a service management system, but essential in multivendor environments.

Data Storage and the Database

Without question, the most important component of any management application is its database and storage facility. These databases are not a collection of flat files or an application created by the management software developer using popular data storage techniques. Those databases are commercial, third-party, industrial-strength software, having support organizations that management software developers and their customers can rely on for help.

The justification for this strong requirement is that management applications have become more complex over time, and the amount of time it takes to write database-type functions detracts from writing useful management functions.

Application developers sometimes speed up a first release by using a free-ware product or by starting out with a simpler form of storage, such as flat files. In the end, these tactics are time consuming and expensive for both developers and customers. Code is written and optimized for the first-generation product, only to be rewritten when a database is put in place. Transition is never smooth and is always disruptive to the installed base. The requirement for a fully functional database exists even in those systems that do not claim any configuration management facilities, because historic information is necessary to perform fault management, capacity planning, and other functions. Thus, *memory resident databases* are not an alternative. (In this context, a memory resident database is one that is reinitialized each time the application is started. It gets its data either from flat files, during the course of operations, or both. Flat files in such systems generally do not store historic information; they store basic configuration information for the memory-resident database to use on initialization, such as which systems are to be polled for state information. The actual state and counter information is not saved in the normal course of operation in such systems.)

In the final analysis, the absence of configuration functions eliminates management software products as real contenders for an integrated service management system in the first place. Our concern is with full-featured systems that can perform configuration functions and verification through ongoing data collection and the analysis of the historic data. The database is central to a service management system because:

- **It determines overall system performance**—In complicated management software, the bottleneck almost always occurs when we attempt to put data in the database or take it out. Ultimately, the upper performance limits of a service management system are determined by how well the database tables and software that access these tables are designed.
- **It defines the types of relationships that can be efficiently set up in the system**—Nobody can describe all the relationships users might want to create with the data in the database. But we can do enough "homework" to define the most frequently used relationships in the system, so that these major relationships will be optimized in the system. To achieve efficient performance on key relationships, such as customer utilization of an important network resource, some data must be duplicated in parts of the database system. There is a cost to this duplication, but generally speaking, this is a good trade-off.
- **It is a major factor in determining how easily the system can be distributed to provide scale and redundancy properties.**
- **It facilitates data migration and upgrade**—Inevitably, new software releases are needed to provide new functions and allow for the expression of additional relationships in the managed system. This is a difficult task, whether you have a custom-built database or you rolled your own data access functions to flat or other file types. Commercial database products have a more fully fleshed out infrastructure to reduce the cost of developing software for migration from one version of the database layout to another. From the user's perspective, this means smoother migrations. From the vendor's perspective, it means producing less custom code, a faster development cycle, and better stability for the resulting product.
- **Third-party components can greatly reduce time to market and overall costs for the development of a fully featured management system**—If a service management system is developed with a third-party database, user interface, and SNMP code for the access layer, the database is the most expensive component.* This has a direct impact on the cost to the customer as well.
- **No standard covers information movement from one database to the next in the way that we specify how data is moved with**

various networking protocols—Because of this lack, many commercial database products have gateways to allow movement of data from one vendor's system to another's. This feature isn't as efficient as standardizing all data in the customer's universe, but it does allow for movement between systems with less custom development by the end users. Database products based on the most widely used third-party software are easier to integrate than products implemented with facilities custom built by the management application developer.

* **It affects the cost of customer deployment. In any large installation, databases require maintenance**—Large organizations tend to standardize on one or two major vendors and train staff on these vendors' software components. If they buy a service management system developed from such components, they won't have to hire or train staff on a new system. In addition, large environments can often obtain discounted licenses directly from the database vendor, saving even more.

    A side effect of this arrangement is that the management software vendor must be willing to sell its product in a modular fashion. One customer may want a product version that incorporates the commercial database and another may not. The product that comes without a database must be designed so that it can be integrated with the customer's licensed database, just as if it had shipped that way from the vendor.

* **It determines how responsive the system is to user customization**—If the system too rigid, end user requests to combine data in ways not foreseen by application developers will be more difficult. The service management system must fit into the service provider's business model. To do so, it must be somewhat flexible in the way it organizes data. This is more easily accomplished with a fully fleshed out data design implemented with "real" database software.

Taken together, the points above are a compelling justification for using a standard database technology. The costs associated with standard technology are potentially significant, but won't outweigh the benefits. The costs attached to standard database technology are:

---

*The database component is likely to be expensive whether one uses a third-party component or attempts the development directly. It is a case of pay now or pay later. Third-party database software does cost; however, the point is that this cost is less in the long run, when looked at from a time-to-market, performance, customer satisfaction, and functional perspective. See the following discussion about database costs.

- **Skills required for engineering**—Far more engineers have general programming skills than have the specialized skills for database design, implementation, and maintenance.
- **Engineering time and time to market**—It takes more time to build a supporting database for service management from scratch than to design rudimentary flat files that can store management data. Time-to-market considerations often push people into the short-term decision against using database technology, only to have to pay a far greater time and resource penalty later when the need for database technology is recognized.
- **Cost of the software and ongoing licensing costs**—Software licensed from a third-party vendor has costs associated with acquisition. These costs vary with the licensing required for each copy of the service management software that uses the database technology. The management software vendor may also have recurring costs for support and upgrades that must be paid to the database supplier.
- **Dependence on a platform and a single vendor**—Most commercial databases offer a number of "standard" interfaces to help insulate code from many of the details of the database. These APIs sit in the architecture on top of the database and are used by the data access layer. They make it somewhat easier for the management software developer to migrate to other databases in response to customer demand or technical needs. In the end, APIs specific to the database vendor often provide better performance, advanced features useful to the developer, or both.

What Belongs in the Database

The types of data that exist in network management applications generally fall into one of the following categories:

- **Configuration data for the application**—Information the application needs for correct operation. Much of it is related to the specific management software installation, such as the locations of default directories for programs the application will use, locations of distributed components, users who have access privileges to the application, and more.
- **Data that the access layers need for communication**—SNMP, particularly SNMPv3, has information about which systems can be contacted, who is allowed access, and what kind of access is allowed.

Other access methods also have information about which users can perform which functions.

* **Operational configuration data**—Data used by the system for its ongoing functions. These include which systems are to be managed, defaults for configuration operations, what data to collect from managed systems, which customers are related to each system, which services are deployed, and more.

* **Historic data**—Data collected from managed devices or synthesized on the management platform from the data collected from managed devices. This historic data includes fault, performance, security, and accounting information. It also includes a history of the configuration of each managed device without which it is impossible to perform other functions. For example, retracing the configuration states of network elements to see if a configuration change made in the past is the source of a network failure.

An ideal system incorporates all these data types, but time and other pressures sometimes force an incremental approach to implementation. In those cases, the database must incorporate, at a minimum, the operational configuration data and the historic data. These two data types are intimately related and, unless both are in the central data repository, it will not be possible to perform the types of analysis essential to a service management system.

Data Management and Archival

Databases provide basic facilities to move data into and out of them, but the service management application must provide those functions that use these facilities to move data in a way that is meaningful for the service management application. A customer of a premium service may require information about a fault or service-level failure. To retrieve this information from the database, we need to retrieve data from devices known to support the customer's specific service. Some environments may be required to hold these data for long time periods, often exceeding 12 months. If we were never to "prune" the database, it would become full of mostly irrelevant information and, ultimately, performance would suffer. Much of the code used for retrieving information based on these relationships can be used for pruning as well. In other words, the same code that extracts the data for monthly output to billing systems can also select a range of these data for deletion. The data can be completely deleted, moved to a backup database, or reformatted for flat file usage.

Many application vendors provide only limited function in this area in early releases; this leads to difficulties when customers deploy and use the software. The only solutions are the user development of specialized code, custom code from the vendor, or operating with reduced capabilities until a new release is available.

Database Access Layer

The database access layer is code developed by the application developers to coordinate all access to the information in the database. It has several important functions since it:

- **Provides a consistent API for applications to isolate them from database details and data migration**—This API is tailor-made for the service management software components and acts as the intermediary between them and the database. This is a practical application of encapsulation, as discussed in Chapter 5. Rumbaugh, et al. describe encapsulation as:

  > *Encapsulation* (also *information hiding*) consists of separating the external aspects of an object, which are accessible to other objects, from the internal implementation details of the object, which are hidden from other objects. Encapsulation prevents a program from becoming so interdependent that a small change has massive ripple effects. The implementation of an object can be changed without affecting the applications that use it [RUMBAUGH].

  Many basic objects in the service management system are realized in the database access software. Analysis software accesses these objects for a variety of reasons: to collect usage data, to change the state from active to not-active, or to enter a failure condition.
- **Optimizes read and write access to the database**—To obtain the best performance from the database, this software may organize the requests for information from or insertion into the database.
- **Coordinates access to data among management system software components**—The coordination function provided by this software is related to optimization. Several modules may request information about an object at the same time, or try to change its state. Rules about who gets to go first might be included at this layer.

System Control and Distribution

If the service management system goes down, it may not be possible to reconstruct the lost data. In environments that perform cost accounting or bill generation, network operators may find themselves unable to provide the information those systems need. Whether the system went down because of overload or hardware failure, the results are the same. Network management vendors who understand the requirements for high-reliability systems have been expanding the distribution facilities of their software for some time.

A service management system should mirror the network's reliability features to ensure continuous service delivery. Distribution can bolster high reliability for the management system: in case of a hardware failure or a loss of connectivity between a service management system component and a section of the network, a duplicate running backup located elsewhere in the network might still have connectivity. Distribution is also one means of improving scale: multiple components of the same type share part of the total load. We want to be able to distribute all components in our management system including databases, analysis modules, transaction control, network access, and user and upstream interfaces.

The system control and messaging component is responsible for distributing the various components, keeping track of their state, allocating work, and sending other messages as needed. This requirement for distribution is one of the compelling reasons for using a commercial database. A great deal of work has already been done to enable the distribution of database technology across many systems and provide online hot backups. Developing these functions from scratch is time consuming and difficult.

Analysis Modules

Analysis modules are where the "brains" of the management system reside. Few systems have a full range of configurable analytic functions, but many have some smarts at the protocol level. What is protocol-level configuration? Although systems may know how to transmit a configuration file, they do not know if the contents of that file are correct for the target device. That is, they can correctly send the data using the selected transport protocol, but they have no idea about whether the configuration information contained in the file is correct. Some have a bit of smarts in this regard, but few take into account the local context of the device or its utilization patterns. Most systems that claim to help with fault management also use protocol-level support—they may be able to receive SNMP notification information and display it on a map or scrolling list, but do not understand a lot beyond this.

Some systems can even assist with root-cause analysis by looking at the notification message data received and suppressing duplicates or notifications of secondary or tertiary importance. More value can be obtained from the notification messages if the management system knows the provisioned status of the failed component. This information can be used to determine the severity of the failure (i.e., failure of a device that is not in use may be less important than the failure of a device serving an important customer). These types of systems are beginning to move from a protocol-only level to something more complete (and thus useful) for operators.

In some implementations, analytic features might be in the database access layer, in the user interface layer, or distributed among various components of the management software architecture. Figure 11.1 shows modular analytic features to highlight the value of keeping them separate at this layer, where other modules can use them conveniently. It is impossible to come up with an exhaustive list of all analytic functions a service management system might reasonably provide, but we can cover some typical examples. Some of these features are included in the general requirements for a service management system, so with a little more background, you'll be able to see how these functions fit into the overall management system.

Fault Management

* **Interface failures**—To correctly convey an interface failure, a system may need to know the configuration status of the interface. For example, does it carry traffic from an important customer? Determining whether an interface has failed or is the result of the failure of some other component, such as a card that has several ports on it, depends on some knowledge of the hardware and software configuration of the system.
* **Service failures**—The notion of service should be locally defined for high-value services. Therefore, the fault management software must

be dynamically reconfigured as these new services are defined, so that they can be effectively monitored.

※ **Failures related to software components (local or remote)**—Some software component relationships can be built into the management software, while others must be configured on site, based on the types of services in use. The basic function can be extended to connect services spanning many systems that can change over time. For effective fault management and service reporting, the management system must be able to respond to these real-time changes in software component relationships and report on them, given its understanding of the desired interdependencies. For example, if we have an application that depends on a database, we will have an application failure if the database fails. Clearly, we want to know this, but of equal importance is the knowledge that the real cause of the application failure—the root cause—was the database failure.

Configuration Management*

※ **Valid configuration based on local rules of the managed system**—Rules for correct configuration are seldom the same from one vendor to the next. Indeed, they can change from one software release for a system to the next. Basic rules can be provided, but the configuration management analysis module itself must be configurable with the rules to apply not only on a per-service basis, but on a per-device basis across a network.

※ **Valid configuration based on utilization**—Some service providers have policies about how high they will let utilization run before they deploy additional resources. Your service management system should permit analytic functions to be configured with these usage policies, so that services are not deployed on systems at or near exhaustion.

※ **Valid configuration based on state**—It makes little sense to configure a service on a device with components that are in a failed state or have evidenced inconsistent behavior in the recent past.

---

*These are yet-to-be-developed features. We lack software that can perform these powerful functions. That said, we still need them if we are going to deploy high-value services.

Accounting and Service Level Management

* **Over- or underutilization**—Over- and underutilization decisions are made at many levels of the network environment. At one end, a network device may discard an incoming packet because that type of packet has exceeded its configured maximum. At the other end, a billing system may add a surcharge because the customer exceeded his guaranteed load by a certain percentage for a certain period of time.

  The service management system does not need to know about the pricing or billing parameters. It *does* need to know what thresholds to look for and over what periods of time to look for them. It then records these events for handoff to the billing system. Suppose you're providing a Web service guaranteed to serve up to some maximum number of pages per hour, and several times during the month a customer exceeded the limit (and you served the pages anyway). At a minimum, you'd want to record these events, and you'd probably also want to know how many pages above the limit were served each time. This information could help you apply different pricing to these over-the-limit pages.

* **Nonconforming service times**—Not all services required 24×7 availability. The service management system must know exactly when the services are supposed to be operational and detect when the service is not available during these times. Information about service violations is handed off to the billing software.

Performance Management

* **Redistribution of work based on load**—Some products move traffic based on observed load, and capturing this information is valuable. A sophisticated service management system might observe load in many network components at once and dynamically reprovision the network to deliver the desired load and performance on each element in the network. The management system has access to all desired parameters and performs calculations based on the configuration and real-time data received.

* **Capacity load limit reached**—Capacity is not always easily determinable from a managed device, nor does performance always degrade in a predictable way. Based on the observed responsiveness of various system components, a service management system might generate an alarm long before performance is compromised.

Security Management

* **Access violations based on configured values**—It's rare for a certain type of access to be universally proscribed. I might allow access to some application servers from outside my local network, but limit this access only to servers on certain networks. Only in the case of access attempts from nonauthorized networks would I need to log an event or send an alarm.

Some of these analyses can be preconfigured in the management software as delivered from the software vendor. In other cases, the types of relationships and the rules to follow with respect to fault and other types of analysis are better configured by the user to reflect local policies. From an operational perspective, the level of configurability for each function can be even more important than the function itself: it's the former that determines how well the system can adapt to the operational environment.

All of the functions that we have described (and many more) depend on the types of relationships supported by the system. In the next section, we take a closer look at relationships that are potentially quite helpful in creating a service management system. We'll see how the requirements explained in Chapter 2, supported by the appropriate relationships, become working features of an integrated system.

Data Relationships and Abstraction

Effective network and service-level management depends far less on management protocols than the heated debates about them would imply. Effective management is primarily based on data relationships and their realization in databases.

I've argued that a flexible system based on a commercial database is of highest importance. It is one piece of the software system that will allow more advanced users to create relationships that better fit their business needs. As we review various relationships, regard them as potentially important inclusions, not as the only valid set. From an operational perspective, these relationships could become a checklist of questions to ask when evaluating software. They're good ways to think about the new high-valued services and the types of reporting management customers will expect.

This discussion once again employs UML to represent the model for our service management system, and the diagrams are somewhat more complex that those we saw in Chapter 5. The diagrams that follow represent one possible expansion of basic concepts into a fuller model. Keep in mind that the level of detail in these drawings has been restricted to just what's necessary for illustrating basic ideas. They do not include additional information that would be requisite if one were to actually build a system from this model. For example, no data types are defined for attributes of the classes, and no parameters have been specified for the methods. It is easy to imagine other useful classes and certainly many more methods and attributes for those classes that have been defined.

Customers and Service Relationships

Customer and service relationships are arguably the most important in a service management system. These relationships are the foundation from which other relationships are made. It is impossible to generate a bill without knowing which customer the bill is for and what specific charges should be associated with them. It is not possible to know what to bill without knowing what resources in the network have performed work to deliver the services and what quantity of work has been performed.

Figure 11.2 shows some of the important classes you'd expect to find in a service management system with facilities for associating customers with services. In the customer class, a few of the attributes have been filled in.

The *customerId* is a unique value that, in a real system, would probably be an integer. Type assignments are omitted from these diagrams because types are not always as straightforward as an integer; including them would bog us down in programming language specifics. Many of the classes in theses diagrams include an ID, because that is a convenient handle for many purposes. As in other places in the system, using an integer ID for instances of objects in our system, as opposed to some character-based string, can be one way to improve performance of the system.

The *CustomerRelationshipList* demonstrates an important concept for service management. Customers can and do have relationships with their own customers, thus making it possible to determine which customers are related to each other. Some systems may even want to track the type of relationship. It's important because the provider may want to create services that let a customer pay for enhanced secure service to its application servers for a select set of *their* customers. In these cases,

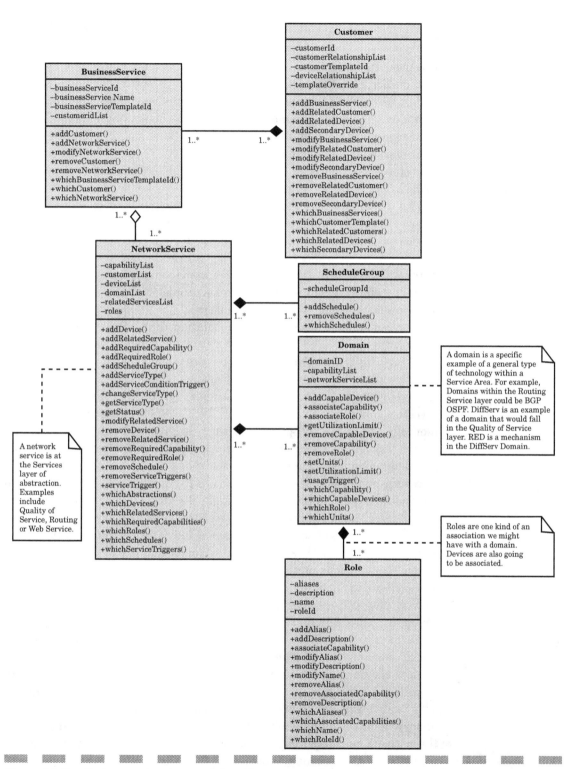

Figure 11.2 Business and customer relationships.

the provider must identify the related customers and track them for information about service interruptions and volume of service provided.

Just as network devices have defaults for the most commonly set values, customers can have common attributes. In this example, *templates* are used to store many defaults. The level templates diagram, Localization of Defaults, shows just some of them. The *customerTemplateId* is an attribute that stores the reference to a customer template from which a given customer's defaults were created.

The *deviceRelationshipList* is critical for our ability to accurately report service reliability and usage. It contains all devices in the network that are used to provide one or more services to the customer. Some of these relationships might be entered by hand, while others are automatically filled in by the service management system. This is true for many attributes in all classes, reflecting the requirement to have the software do as much work as possible while still allowing users to override software decisions or fill in information that is not determinable by the software.

Using the *templateOverride* attribute shows an example of how defaults might be overridden in our system. Depending on how a system is implemented, you might find a number of different *templateOverride* attributes or just one complex list. These types of attributes and the methods necessary to operate on them allow us to use templates and control aspects of the system from appropriate defaults while allowing for custom changes on a per-customer basis.

Notice that the type of association in Figure 11.2 is shown as a composition (the solid diamond at the end). Recalling our earlier discussion, this implies that Customer objects are made up of several different components, as opposed to all of the same kind. That is the intention here, although not all of these potential associations are shown in the diagram. The first four methods shown in the *Customer* class are for adding associations with three different classes: *BusinessService*, *Customer*, and *Device*. The rest are simple examples that modify, delete, or show existing relationships. In a real system, there would be many more methods.

Notice, too, that there are several methods for secondary devices (e.g., *addSecondaryDevice*). The idea here is to let the system know which devices are backups, so that it can send notifications to operational staff to convey whether a primary or backup (secondary) device has failed. This is also relevant from a billing perspective, since the failure of a secondary device does not imply a service loss. Indeed, depending on how the network and services are configured, it might require the failure of all primary and secondary devices before the service ceases operation.

This information about the state of a particular device will be found elsewhere in the system. Analytic software takes information from the Customer and various devices classes, along with recorded information about state and utilization, and determines if there were service interruptions during a billing period. This same software could also be crafted to help send an alarm in cases when a specific service has been impaired or undergone a complete failure. Type of software can help service providers better meet their obligations to customers, not only to keep services running, but also to notify them when there is a problem.

In our diagram, a business service is composed of one or more network services.* Network services alone define our business service, which is why it is shown as an aggregation. The system is defined so that many business services may be associated with a customer and many customers can use a business service. Although most service providers attempt to keep the number of business services they offer to a minimum to simplify management and billing, I have included a *businessServiceTemplateId* as an attribute in the *BusinessService* class. This allows operators to change the characteristics of a business service at the template level, where the change is applied to all customers of that service.

A business service is expressed in a way that customers understand: a certain quantity of a particular service over a specific period of time. Thus, a business Web hosting service will serve a certain number of pages per hour, probably with a specified maximum latency, through the service providers network. The service might also include secure connections for sensitive information on the Web server or application servers used by the Web server.

Now, let's isolate some specific relationships that belong in a service management system. Remember that network services are complicated and not all logical combinations of add, modify, delete, and show methods are included in Figure 11.2.

* **Devices are required to deliver services**—Each network service will have one or more devices that have been identified as being able to deliver the specific service. We will examine some of the qualifiers that determine whether a device should be added to the *deviceList* by the *addDevice* method. In general, if a device is supporting the service

---

In this diagram all cardinalities have been entered only as placeholders. They have all been entered as 1.. at both ends of the association lines. A refinement of this class diagram would add detail and perhaps change some of these cardinalities. They serve our current expository purpose well for now.

or services on which the high-valued service depends, we include it in the device list.

- ▨ **Some services require the proper functioning of other services**—A Web service that only allows access over secure lines is obviously related to another service that defines those security properties. The ability to associate services is fundamental to the delivery of high-value services and for their billing. If one network service that is part of a business service fails, the operator may not be able to claim that he has delivered the business service at all within that time frame. The *addRelatedService* method adds related services to the relatedServices list.

- ▨ **A capability in our system is used similarly to the way SNMP-CONF uses it**—It represents some set of objects that collectively indicates an ability to do a specific kind of work—deliver a service. An important distinction for our system is that it's designed to be access-method agnostic. In other words, the model described here can support systems that use CLIs, SNMP, and other access technologies as well. As a result, in this system, we are not (yet) dealing with SNMP OIDs or CLI commands for the description of capability in this class. How we associate capabilities with devices and access methods is described in the Object Relationships diagram.

- ▨ **A network service can be expressed in terms of the capabilities required to perform the service**—For delivering a Web service, one vital capability is a Web server. Capabilities are one way to restrict the addition of a device to the *deviceList*: if a device does not have the required capabilities, then it cannot support the service.

- ▨ **Roles are used here in the same fashion as in the SNMPCONF technology**—Roles are one way to determine which network resources and portions thereof should be configured to deliver a specific service. A management system might associate a secure service role to specific interfaces throughout the network to deliver a secure Web service. Management software then uses those associations to determine what type of configuration and monitoring information is necessary for those elements that support the service. Sometimes it takes a combination of roles to determine the elements on which a specific service is to be offered. For this reason, we have a method, *addRequiredRole*, that can add as many roles to our roles attribute as required. A fuller exposition of this system would almost certainly have methods and attributes that deal with element types, as defined in the chapter on SNMPCONF technology. Whether one uses SNMP-CONF or not, the idea that we might only want to use certain types of elements, such as Fast Ethernet interfaces, for certain services is a reasonable one.

* **Business services, especially complex ones, might not require the same operation 24 hours a day, 7 days a week, 365 days a year**—For this reason, several network services may be defined to support business services with schedule variants. One might only run on holidays or weekends, while another might have a complex set of schedules that turn the service on or off. These complex schedules are defined in a separate class, the *ScheduleGroup*. Each of these business services might have several different *NetworkService* instances associated with them, each with a *ScheduleGroup* that controls the hours of its operation.

* **The addServiceConditionTrigger and serviceTrigger methods represent an important binding of state and usage information with the configuration information that is part of this class hierarchy**—A service trigger is the specification of some state or condition that merits special attention. When this condition is reached, as monitored by the software, the serviceTrigger method sends messages to the relevant software modules for logging and/or user notification.

* **There are two types of services in most network environments**—The familiar routing services do not (usually) directly produce revenue. They are a foundational element on which other services can be built, and a prerequisite to proper operation. The second type of service does produce revenue by delivering something of value to a customer. Understanding the relationships of these service types helps network operators react to an event appropriately. For example, many customers are affected in the event of a routing failure, although none of them have specifically purchased the routing service. The several *ServiceType* methods in the *NetworkService* class are included so that software can be written to determine which type of service is an instance of the *NetworkService*.

* **The *getStatus* method allows the system to determine and log the status of the service at any point**—This is a useful tool if one has to demonstrate the health of the service over time.

You may have noticed that, in Figure 11.2, domains are associated with network services. This is one way to connect the abstraction of a network service to increasingly specific management information. Several important associations are implied by the *Domain* class, its methods, and its connection to other classes in the diagram:

■ **Devices are associated with specific domains based on their ability to support those domain types**—So, whereas at the network service level we only care about *which* devices can deliver the service, at the domain level we refine device distinctions in terms of *the way* they deliver the service. A device is associated to a domain by assessing its capabilities. Notice that we *associateCapability* with a domain, and we use the *addCapableDevice* method to associate specific devices with that capability. In this case, as elsewhere, software must be written to realize these methods. The trick to getting this system to work correctly is not to hard code the relationships of devices and capabilities. Rather, use flexible code that allows these associations to be modified as new software and hardware products are released.

■ **Usage limits may be established for a domain and monitored on an aggregate rather than per-device basis**—Customers are not as concerned about usage on an individual device as they are about the total amount of work done on their behalf. The exception, of course, is when a device fails entirely or falls short of performance expectations. The *setUtilizationLimit* method allows us to limit the total amount of work done in a specific domain used by a service over a specific period of time. The customer association with an instance of a *NetworkService* is important to our ability to see this utilization and other important state information on a customer basis. One of the problems with much of the management technology in use today is that the units of measurement are not always made clear. With the *setUnits* method, the system can set specific units to monitor and report. The *usageTrigger* method sends the proper messages and logs information when a limit has been reached.

■ **Roles are also associated with domains**—Once a domain is associated with a network service, the service can, if desired, associate the roles that are associated with the domain, thereby providing a global view of all roles associated with a service.

Management Objects

As we've established, systems that can handle differences in technology and access methods are more costly to develop; however, current conditions in the industry require that we do so. Figure 11.3 makes the point concretely by illustrating some of the additional complexity that typifies the development of such a system.

This diagram represents one potential set of
relationships between management aggregates,
management objects, capabilities, and domains.
Note that capabilities can only exist as a part of
a domain when they are below the domain level
of abstraction.

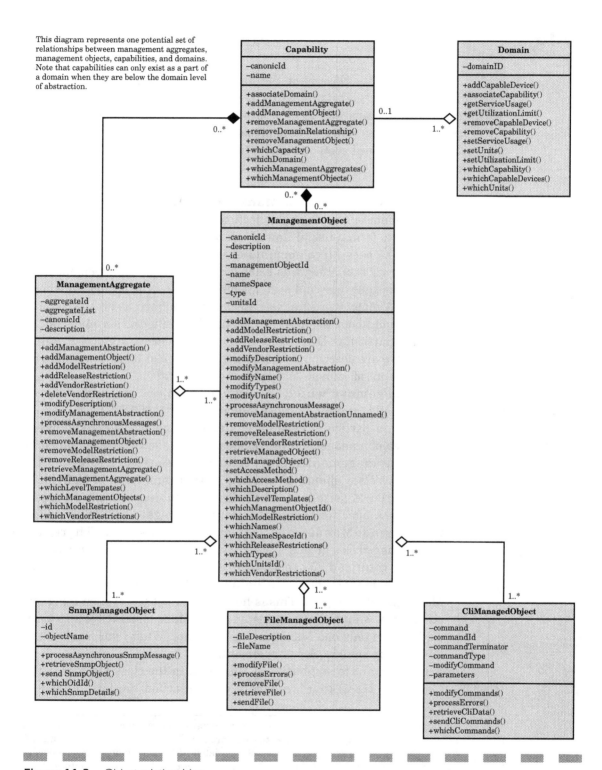

Figure 11.3 Object relationships.

The Capability class is at the top of this diagram because it is one way to connect this picture with the rest of the system. We have already discussed domains and their relationship to capabilities; here we connect capabilities with specific objects that can realize the capability's function. Relevant methods in the *Capability* class for this relationship are:

- **The add/remove/which ManagementAggregate methods**—A management aggregate is a collection of management objects bundled together to form a useful unit of information. They also make it more efficient to pass large amounts of information around the system. Although no real standards exist for this concept, management software in both devices and systems already has some of these methods built in. A management aggregate could be defined as all the counter information about an interface, or all the configuration information about a particular BGP peer. In the management protocol-agnostic system we are describing, these aggregations can be custom-made by the users out of individual management objects.
- **The add/remove/which ManagementObject methods**—These methods add, remove, and list individual management objects associated with a particular capability. ManagementObjects are the building blocks of management aggregates, and any software that supports the addManagementAggregate or addManagementObject in a Capability class should be written so that duplicates of individual objects are not permitted. In particular, you do not want an object added to a specific capability via an aggregate and then added with the addManagementObject method as well. There are many other places in this system where similar kinds of error checking are required. Note that in the diagram, an instance of a ManagementObject can be associated with more than one ManagementAggregate. This is how I would put the system together; others may want to limit an instance of a ManagementObject to association with only one ManagementAggregate. What I suggest is more flexible, but more work to implement.*
- It is important to emphasize at this point in the class hierarchy (the *Capability* class) that we have not yet reached the level of access

---

A close look at the cardinalities shows that they have been specified so that there is a great deal of flexibility, for example the numerous 0.. examples. Is it likely that an instance of a *ManagementObject* that is not associated with a Capability would be useful, but most of the time, I would expect a *ManagementObject* to be associated with at least one Capability.

method–specific objects—those that are defined in MIB Modules or referenced by specific CLI commands. This important mapping is made in the *ManagementAggregate* and *ManagementObject* classes. This extra mapping step is needed to ensure canonic representation of management information for the rest of the management system to use. Regardless of the information's source and how it was retrieved, all data used by the system is of the same form. Namely, they are instances of the *ManagementObject* class, or aggregates of instances of the *ManagementObject* class (in the case of instances of the *ManagementAggregate* class). An important attribute in the *ManagementObject* class is the *managementObjectId* attribute. This is not an SNMP OID: it is a unique integer value assigned to every management object class in the system, regardless of the name space in which the object is realized. The fact that the value is a single integer has a side benefit—it's much more efficient for the software to deal with than the ASCII strings one would use with a CLI or the dotted OID form used by SNMP.

* The *name* attribute is a textual handle that should make sense to users.

* The *nameSpace* attribute indicates the source of the managed object. SNMP and CLI are two examples.

* The *type* attribute contains information about the type of management information an instance of this class carries; that is, fault or configuration. Some objects may have several values. For example, some management objects are important for both fault and accounting management. By having this indicator as an attribute of the management object, we can then identify all fault or configuration information for a system, a device, or part of a device.

* Just as the *ManagementAggregate* and *ManagementObject* classes are designed to normalize management information that comes from different access methods,* the unitsId attribute in the management object normalizes—where it can—the specific metric appropriate to the data in the management objects. This is particularly valuable for counter objects. Note that unitsId would be an integer value in a real implementation, instead of a name like "kilobytes." That's because a unit is actually a complex object in itself, and this ID is intended to point to an instance of a *Unit* class. For example, a unit might have a

---

*The *ManagementAggregate* and *ManagementObject* are key to the normalization process. However the *CanonicId* class presented shortly is another essential "glue" element in the system that helps make it protocol and name space agnostic.

general description and an indication of what type of integer it is, such as floating point. The integer value for units used with management objects is designed to help the system move data around as efficiently as possible.

The *ManagementObject* class is an important contributor to the system's ability to localize management data from abstract to specific data types (refer to Chapter 5). Because different vendors implement features, even those covered by standards, in different ways, how do we localize the management interaction so that it fits each managed device and at the same time maps to some higher-level behavior that is meaningful to network operators and customers? The *add/modify/remove/which* set of restriction methods address this problem.

Two aspects of the problem relate to determining if a device supports a particular management object. In one, the object derives from a private space such as a CLI command. In the other, it derives from a standard name space such as the IETF. In the case of management objects defined for a specific name space, such as that of a specific vendor, it's reasonable to assume that only products by that vendor will support the management object. (You'll see when we look at the device hierarchy that the *Device* class contains a number of attributes to help us with this aspect of localization.) Thus, the rule for private objects is: all devices are assumed not to support a private management object unless they come from the same vendor name space. Even then, not all models and releases are equal. The *addModelRestriction* and *addReleaseRestriction* methods are used to identify whether support is restricted to a subset of the vendor's products.

There is an implicit hierarchy for these methods. It starts with the vendor, continues to the model, and ends with release. To indicate that only model #1 and model #2 products from vendor A support a particular management object, we must add restrictions for those models. Without restrictions, all releases of these models are assumed to support the management object. In the case of a new object that first appears in a subsequent release of these models, we would have to add a release restriction specifying releases that support the new management object.

What about object support in a standard name space? Management software vendors have been refining techniques to make this determination in the SNMP standard name space for many years. The question is, once I know that a particular vendor, model, and release support an object, how do I integrate it with my service management system? Well, if the object is in a standard name space, such as the SNMP name space, one reasonable approach is to assume inclusion by all vendors that support a specific capability. For example, you would probably not look for standard database objects on a router. The alternative is to design the system so that the common name space is considered unrestricted until vendor-, model-, and release-specific restrictions are filled in. This would work in the same way described for private name space objects. If vendor A supports a standard MIB Object, and no further model or release restrictions have been declared, it's assumed that all models and releases support that MIB Object. This type of design decision is always hard unless it can be determined which approach will be most efficient for the users and the system itself, based on the most likely set of equipment to be managed by the software. If the equipment that is most likely to be managed always supports the interesting standard MIB objects, then it makes most sense to assume support as the default, subject to being overwritten by restriction methods.

Another interesting set of methods causes messages to be sent to network devices and processes asynchronous messages from managed devices that support a specific instance of a *ManagementObject* class:

- **sendManagedObject**—This method causes the system to send the object to selected network elements. Because the *setAccessMethod* has already set the type of access method, the *ManagementObject* will know to what type(s) of objects to send this message. In our diagram, these types would be instances of the *SnmpManagedObject*, *FileManagedObject*, or *CliManagedObject* classes. Each of these classes knows the details of how to interact with managed devices in the protocol-appropriate fashion.
- **The *whichAccessMethod* is provided so that other elements of the system may learn the access method used for communication**—An instance of *ManagementObject* uses one access method, as indicated by the cardinalities in the diagram. Note however that many individual *SnmpManagedObject* instances or *CliManagedObject* instances could comprise a single *ManagementObject*. It is also reasonable that an instance of an *SnmpManagedObject* might be used by more than one instance of a *ManagementObject* although this does complicate other portions of the system.
- **The *setAccessMethod* allows the system to set the access method(s) used by instances of the *ManagementObject* class**— Once set, it generally remains constant over time.
- **The *processAsynchronousMessage* method receives error messages and SNMP NOTIFICATION-based information from the**

access-specific classes—This information is passed to other parts of the system for additional processing.

* **When the system needs information from managed devices, the *retrieveManagedObject* method is used**—This method in turn uses appropriate instances of the access method-specific classes.

Before we leave the discussion of object relationships, we need to take a closer look at an attribute that is present in many of the classes in this diagram, the *canonicId*. In each of these classes, this attribute points to a *CanonicId* class (not present in the diagram). This tool is used to help make our system protocol agnostic. A single instance of a *CanonicId* class may be pointed to by several instances of the ManagementObject or *ManagementAggregate*. We could have different management objects for each access method (e.g., SNMP or CLI). Indeed we might even have several management objects using the same protocol, such as SNMP are really semantically the same. This is especially likely when we have vendors that have MIB Objects in their private name space that really mean the same thing as those available from a standards body like the IETF. A single instance of a *ManagedObject* will be associated with only one instance of a *CanonicId* class.

Device and Service Relationships

Devices can be thought of as hardware, service-delivery platforms, or resources. The Distributed Management Task Force has generated a number of models for network elements. Figure 11.4 is a generic one designed to fit into an overall network services model and does not incorporate CIM terms, but the two approaches are not mutually exclusive.

Using device relationships, we focus on the association of service elements and the roles assigned to them, along with their capabilities and capacities to do work. Once made, these associations allow the system to map from the abstract services described in the business and customer relationships diagram, through other layers of abstraction, down to the access-specific objects found on each service element. The next dimension of localization to address is mapping services to specific devices by means of each device's capabilities and the roles assigned to the parts of the device, as represented by the *ServiceElement* class.

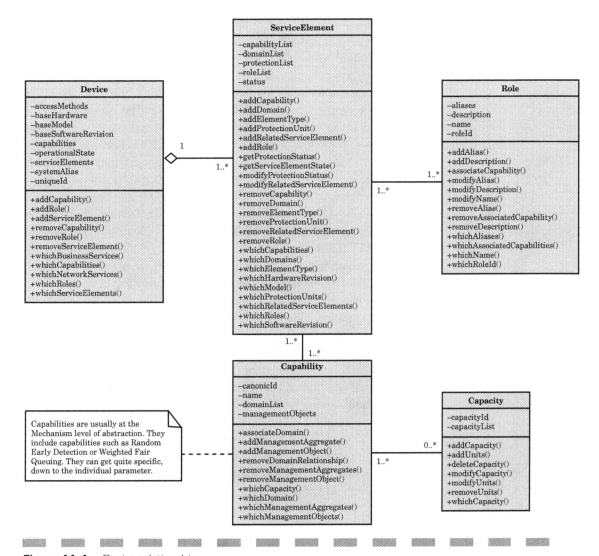

Figure 11.4 *Device relationships.*

A device can be thought of as consisting of service elements such as disks, interfaces, Web servers, routing functions, and the like. That relationship is shown by the association of the *Device* and *ServiceElement* classes in the diagram.

Many of the attributes in the *Device* class can be of assistance in helping the service management system determine what types of services and management objects are supported in that device.

In a real implementation, more methods are associated with a device. Here, we focus on those that help flesh out major relationships in the service management system we are describing.

- **The add/remove/which *ServiceElement* methods are used by the system to control the association of service elements with devices**—In some cases, associations may be manually performed and, in others, they are assisted by software. SNMP-based management applications capable of managing these associations are already commercially available. As we noted previously, a simple way to do this is by checking to see if specific SNMP MIB Object instances exist on a device and, if they do, the device is deemed to have an instance of the element inferred by the MIB Object. This technique is extended here to identify if a device supports a specific capability.
- **The add/remove/which *Capability* methods control the association of capabilities with a managed device**—For the most part, capabilities are selected from among those specifically associated with each service element, as shown in the diagram.
- **The methods associated with roles operate in a similar fashion**—Specifically, roles are usually assigned with specific instances of service elements in each managed device. By making these roles visible at the device level, we can conveniently determine what roles a device has been assigned without having to search through each service element.

The *ServiceElement* class at the center of the diagram is where most of the interesting associations are found. This is where roles and capabilities come together so that managers can ascertain if a specific device can support a particular business service.

- **Methods and attributes for capabilities are provided so that we can map specific elements to specific capabilities**—Mapping lets us associate instances of the *ServiceElement* class in a device with the management objects relevant to that capability.
- **The add/remove/which *Role* methods and *roleList* attribute provide mechanisms for controlling and showing the associations of roles with service elements**—The *whichRoles* method shows the roles currently assigned to a service element, and its *add* and *remove* counterparts perform their respective functions. *Which* can be used to determine whether a service element with a particular set of capabilities should be assigned to the delivery of a service, as determined by a specific role or combination of roles.

- The add/remove/which *ElementType* methods enable the system to group similar *ServiceElements* together in a system or network that might not have been already associated through capabilities or roles—Fast Ethernet is an example of an element type that would be useful in the process of policy localization.

Related Elements

We've said that reliable networks and the services built on them pool equipment to act as backups in case of failure. In some cases, a device or service element within a device might be set aside as a backup and not used at all until there is a failure. Setting up the network this way is one thing, but constructing the management software to be aware of these important relationships takes some additional work. The *RelatedServiceElement* and *ProtectionUnit* methods and related attributes help get it done.

For the purpose of this design, a related element is one assigned to work with another service element in one way or another. Each service element may have many other service elements related to it for redundancy and other purposes. If a management application knows the state of all the relationships and the service elements to which the relationships point, it can give a truer picture of the real state of a network service and more accurately account for outages when they do occur. Similarly, it will not deem the service down if a *protectionUnit**—one type of related element—is still functional. The *getProtectionStatus* method provides information about the configured state of the service element, with reference to whether or not it's a backup. It is different from the *getServiceElementState* method, which shows the state of the service element at a particular moment.

This level of complexity is required to guarantee and accurately report on premium paid service levels. Figure 11.4 represents the Role-to-ServiceElement and Capability-to-ServiceElement relationships with a simple association, to avoid any implication of a whole–part relationship.

---

*A protection unit is commonly understood to be a system or portion of a system that "backs-up" another service element. For example, redundant disks are sometimes in place to ensure that, if there is a failure, service and data are not lost.

Capacity

The *Capacity* class is key to creating a system that can provide reliable high-quality network services. This facility makes it possible for the management system to tell when a specific *ServiceElement* element is nearing a resource limit. Without it, we could inadvertently add load on a *ServiceElement* element only to discover that it was unable to carry the load during peak utilization periods. Similar monitoring of management objects that contain information about *ServiceElement* resource consumption and performance will help you avoid a number of related pitfalls:

- **Poor anticipation of service failures because of performance problems as *ServiceElements* approach or exceed capacity.**
- **Poor anticipation of needed upgrades, or else spending resources on unneeded upgrades**—This may be caused by insufficient information about the relationship of service load on each *ServiceElement*.
- **Inadvertent overprovisioning of the network**—This may be caused by inadequate information in the provisioning system about the resource requirements for a new service, or about the current utilization state of each *ServiceElement* needed for service delivery. Akin to this overprovisioning is the problem of service failure during peak demand times. Good provisioning determinations are only possible if we collect data on the configured status of each *ServiceElement* based on time of day, and make correlations to those services that demand resources and impact on the network.

A look at the *Capacity* class reveals that methods are in place to add, delete, and modify an instance of the *Capacity* class. This type of control is mandatory, since most network elements do not have any reliable way to know how much work an instance of a *ServiceElement* can perform. We recognize that the performance of network elements does not often change in a linear fashion as they approach capacity limits, and therefore network administrators may wish to change the capacity value for systems to somewhat less than the rated maximum. On the other hand, it's easy enough to imagine a more sophisticated *Capacity* class, with thresholds set below capacity, sending tiered messages as the amount of capacity expended over the threshold grows.

Keep in mind that some network providers regularly reserve capacity not only to avoid overprovisioning but also to help the network function at better performance levels when network elements fail and work is transferred to backup systems.

Creating Defaults

We come around to this point every so often, because it underlies so much of what service management systems do. If a management system requires operators to configure every parameter for every managed device under its configuration control, the benefits ascribed thus far to service management would not be realized. Managers would have to go back to the greybeards who knew the details of each of the technologies and how they vary from vendor to vendor and model to model. They would have to learn which management objects for each system, technology, vendor, model, and release are appropriate to a given system with a specific configuration. By now, that should be a fearsome prospect. But it also may be a familiar one: this is the situation many network operators find themselves in today.

The right service management system will provide a facility for the use of defaults localized to a managed device based on:

- Appropriate layer of abstraction and technology
- Name space (e.g., vendor-specific or standard, such as the IETF)
- Access methods
- Vendor
- Model
- Release

Methods for the localization of defaults exist in the *MechanismTemplate* class shown in Figure 11.5. A more full-fleshed-out diagram might have them in other classes as well.

Figure 11.5 also includes a *CustomerTemplate* class, because management objects are not the only data elements in our system that benefit from defaults. Part of the worth of a service management system lies in how well it facilitates the deployment of new services and how easily it accommodates the addition of new customers. Frequently, more than one default can apply to classes of customers captured in the *CustomerTemplate* class.

The *TemplateList* class conveniently creates groups of related templates that can further assist in deploying new services or managing existing services.

Figure 11.5
Localization of
defaults.

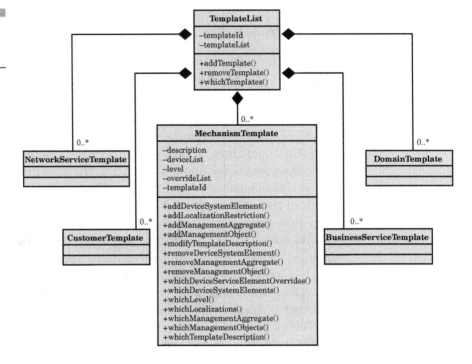

MechanismTemplates are the means by which specific collections of management objects and aggregates are associated. They are localized to specific devices through a number of attributes and methods:

▨ **The *addDeviceSystemElement* method is used to build associations between service elements and their defaults**—This makes it easier for a single change to be applied to a set of defaults and propagated to all the devices that use that set. The *overrideList* attribute lists default devices and objects that have been overridden in the system. A need for exceptions always exists, and allowing for them in your design makes it easier for operational personnel to cover the majority of systems and at the same time control and monitor the exceptions.

▨ **The *addLocalizationRestriction* method is used to associate a specific *MechanismTemplate* with a class of systems based on vendor, release/revision, and model**—We do not have to worry about access method or name space here, because those properties are defined as a part of each *ManagementObject*. As a result, you can have a template that includes multiple name spaces and access methods (most devices have multiple access methods). Some configuration parameters may only be accessible via a CLI, and other counters are best retrieved via SNMP. Each instance of the *ManagementObject* class has methods for associating templates. (See the *whichLevelTemplates* method in the Object Relationships diagram.)

* **The add/remove/which *ManagementObject* and *Aggregate* methods give the system its ability to manage which objects and aggregations are members of a specific template.**

The preceding discussion and diagrams do not represent an entire design for a service management system. Our focus has been on some of the most important relationships that must be supported by a system that purports to manage high-value services—relationships which, for the most part, haven't appear in common IP network management systems to date.

CHAPTER 12

A Complete System

In this chapter, we tie together everything we've learned by adding some important topics in the context of an operational service management system. Based on our understanding of what the edge of a network is and the special requirements for high-value service management, we have reviewed technologies and design issues for service management applications. We have also examined the software that executes in managed devices, including many of the important types of data that must be configured and monitored in the managed devices, and some of the technologies that can assist (such as the [PM] described in Chapter 7). Four additional topics must be discussed to round out a complete service management system:

- An examination of how information might flow from the service management system to the managed devices and back.
- The important role of topology information in a service management system.
- Security considerations for a service management system.
- Customization requirements for the service management system.

In Figure 12.1, we see the two main elements of a service management system: the service management system (applications) itself and the managed devices that are controlled by the service management system. The diagram shows information important to the proper functioning of the system and how it can reasonably be expected to flow from one component of the system to the next—such as, configuration data that moves from the service management system to the managed devices. There is no explicit ordering to the information flows in the diagram, but the following steps comprise one possible flow that would make sense if you were configuring a service on managed elements, monitoring it for compliance with service-level agreements, and outputting the results to a billing system.

1. Input of information about devices, access methods, and the like. Some of this information could come via a discovery process run by the service management system; some might be entered by hand or come from other systems.
2. The policies that govern the operation of the system and the services that are deployed on the network must be available to the service management system. This information could come from manual input directly to the service management system. No matter which system is used for policy information (assuming that a policy-based

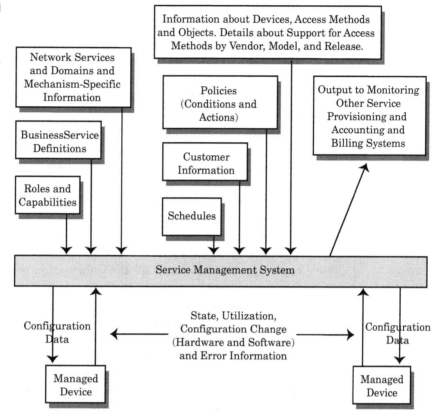

Figure 12.1
A complete service
management system,

approach is used), it must be input to the system. Some may prefer to use an external system, which is fine, although it makes more sense to use the single service management system as the policy repository, because the data must reside there anyway.

3. Information about business services and customers can be input as a separate step or along with the policy information, because they are related. This information can be input by hand or retrieved from a customer database or other source. Part of this business service definition and customer information will include details of the schedules required to support the specific customer services. Different customers may have the same basic service, yet some may pay for premium services 24 hours a day, seven days a week, while others only want that service during their normal business hours.

4. Role and capability information can come from a number of sources. The roles must always be created and added to the service management system. They are then applied by the network operators on

each machine and on the elements in the machine. It is easy to imagine how some roles could be algorithmically assigned to some service elements to ease the load. Remember, though, that roles are often not algorithmically determinable and might therefore be applied at least once as each service element is added. The capability information is not that much different: in some cases, it can be determined by the service management software without too much human intervention. In other cases, capability information must be input to the system either manually or from some other system. Note that a lot of this information can be "discovered" by the service management system, as we will see when we discuss topology and discovery in the next section. In this case, capability information might be included with the configuration change data, which moves from the managed devices to the service management system in the diagram.

5. The final set of data that must be placed in the system before we can begin operation contains all the information needed to properly configure each service element in the network to deliver all the services that will be deployed. This is where the experts really come in, those who know the details of each parameter value to be set for every type of device in the network. There is no shortcut to this, unfortunately. The good news is that once they are defined, the system can hide these grisly details from normal operators. Additionally, if a change is required, the expert can make it once and the service management system can deploy this change everywhere it is needed.

6. Using the above five steps, we have the data necessary to deploy services in our network. The next step allows the service management system to send configuration data to the managed devices, as shown in the diagram.

7. The managed devices collect state and utilization data and, in some cases, send information back asynchronously. The majority of state and utilization data is collected from each device via regular polling by the service management system.

8. The service management system consolidates the data from the previous step and makes reports available. It will also make the data available for export to billing systems.

This seemingly obvious sequence (now that we have gone through so much detail) shows the power of the system we have described and how it could be deployed in a real network. Even with all this power, there is one important type of data we have not discussed in too much detail: topology.

The Role of Topology and Discovery

Topology data is not included in Figure 12.1, although it is information that spans multiple devices and provides context for much of the other information.

Network management systems have included facilities for the discovery of network topology information for many years. Although various techniques have been used to accomplish this function, the goals have often been similar. Topology and discovery have been used to:

- Find out what network elements exist in a network.
- Learn details of the elements in each network device. These details are then used to populate network maps that can show the status of the devices and the elements, identify what the management device can do (its capabilities), and determine what management objects will be used to collect data for fault and other types of reporting.
- Establish connectivity at different layers of the network stack. The IP layer is the most common.
- Act as a baseline for observing connectivity and performance characteristics over time.

Some network operators believe in maintaining an authoritative central repository. This database contains the desired network configuration, and what discovery software sees should be measured against it. Where the two differ, steps can be taken to reconcile them. Networks that do not employ this level of discipline use the discovery process to keep track of new device deployments and configuration changes.

When analyzing a service management system, keep these two provisos in mind. First, all of the discovery and topology features are also essential for a service management system. Second, because many devices do not know about backup and standby status, not to mention other important attributes we've discussed, discovery should be used in the verification mode. Services should not be turned on automatically on a device or service element until they have been expressly configured by the management system. These systems and services will then be monitored by the network operators.

The topology and discovery components of a service management system may exist as part of the analytic modules or as a separate specialized module. Regardless of the implementation details, the discovery system will also have to learn additional information, and some of it will be found in those devices that implement SNMPCONF. In devices that do not, look for other attributes that may point to the same information, such as roles and capabilities.*

Roles, as defined in the SNMPCONF context and illustrated in our diagrams on relationships, should be collected during the discovery process. Most newly installed systems will not have any roles assigned, but should still be checked during discovery. Each time the discovery runs, roles should be reinvestigated to ensure that the management system and managed devices in the network are in sync. This reinvestigation should be performed even though notification messages are supposed to be sent from device to system whenever a role is changed. It is not safe to assume that all role changes are authorized or get delivered. Whether you use SNMPCONF or some other technology for the collection and distribution of role information, the names used for roles should be carefully thought out before beginning their use to reduce the possibility that you must make mass changes as the network evolves.

Capabilities must undergo the same scrutiny as roles. One of the first things to do when there's a new system in the network is to configure the Capabilities Override Table.† In the absence of SNMPCONF technology on managed devices, solutions are still possible. You will recall the discussion we had earlier about discovering a device's set of capabilities. By checking for the existence of specific SNMP OIDs in a managed device in the discovery code, one can simulate some of the facilities found in the Capabilities Tables in the SNMPCONF PM. Similarly, devices that do not implement the PM will not have information about roles; substitutes can help. For example, many networks assign information to the *ifDescr* object [RFC2863] about where interfaces are pointing—roles could be added to this text string.

```
ifDescr OBJECT-TYPE
    SYNTAX        DisplayString (SIZE (0..255))
    MAX-ACCESS    read-only
    STATUS        current
    DESCRIPTION
            "A textual string containing information about the
```

---

*Since SNMPCONF is not yet available in any meaningful way, the alternatives will have to be used until it is or some other technology with similar capabilities is deployed.

†The Capabilities Override Table in the Policy Module lets the network operator prevent the utilization of a network resource in the delivery of a service for which it has the required set of capabilities. The operator may wish to keep certain resources in the network in reserve for a number of reasons. One common reason is to have extra capacity in reserve in the case of failure in one or more network elements. In systems that do not employ SNMPCONF technology, this feature may have to be implemented by hand or through some other approach.

```
                    interface.  This string should include the name of
                    the manufacturer, the product name and the version of
                    the interface hardware/software."
    ::= { ifEntry 2 }
```

What do we do about noninterface element types, such as processes? We could use processes, with names that include roles, to help identify what types of roles they were performing. As you can see, the lack of a standard place for role and capability definition adds complexity to the service management software, because it now has to look in a number of different places to find this information. More important, it must know how to "read" the data returned from these objects to pull the role information out.

One other alternative is not to place any role or capability information in the managed device itself but rather to add this, as an attribute of the device and the service elements it contains, inside the service management application. This entails more work than if the underlying infrastructure supports the notion of roles and capabilities, regardless of the technology used.

Some services may require a great deal of interaction with topology information for proper provisioning. Some of this information may be directly available from a discovery process or other method, but other information (such as backup paths and cost of paths) might be more difficult to obtain. However, all of this information is necessary for proper service operation and verification.

Paths

The service management system does not have be a routing engine, but it may need to know about priority paths or paths with low latency to configure the type of service required in the service-level agreement:

* **Backup paths**—Some services require high reliability, and it may be helpful to know the available capacity of both primary and likely secondary paths through the network when service is configured for a customer. There is little value in having a backup path if it is oversubscribed when it is needed most.
* **Cost of path**—Not all paths through the network cost the same. Certain services may merit high-cost paths, while others do not. The latter may have to suffer degraded performance in various conditions, such as when the primary path fails.

The service management system must also be aware of changes in the path of services used by customers. It uses that information as a basis for deciding how and when to access devices on the alternate paths, to verify service levels on these alternate paths, and to continue to collect accounting and billing data.

Security Requirements

A critical responsibility for a service management system is to ensure that the access it grants is consistent with the security configured on each of the managed elements. To the extent that it is not, the variations should be by design rather than accident.

Figure 12.2
Security in a service
management system.

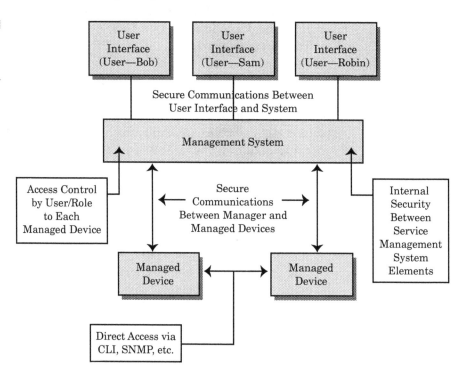

In Figure 12.2, the service management system is depicted as a single block. Given our discussion on the importance of distribution, this might seem odd. The single block is purely conceptual and, in reality, contains many components spread over many systems in a wide geographic area. Whenever a distributed system is installed, the mechanisms for communication between components should be appropriately secured, regardless of the security implemented between the network services layer and the managed elements.

Access to the Management System

If the user interface of the service management system can be executed on remote systems, then those communications must also be secured. The mechanism for doing this is up to the management system's developers. A reasonable choice for such communications is SSH®. This technology can be applied whether using a CLI or an advanced interface in which Java applications communicate with the management system using TCP.

Once a user accesses the management system from the user interface, the upstream interfaces and the system access control component must perform several functions (Figure 11.1):

- **User verification and access control to information in the management system**—Multiuser computer systems not only control who can access a system but what information they can read, create, and modify. A service management system should exercise the same control over its own information. This is in addition to services provided by the operating environment. Although the security services of the operating environment are distinct from those of the service management system, the familiar model of users, groups, and access privileges is useful for both. It is quite flexible enough to let service management systems control what each user can see, create, and modify by user and role. Some users might be allowed to see all customer data, while others may access only low-level statistics.

  There are two other benefits to this approach: first, it suits the way many network operational staffs are organized. Groups (people with specific roles) have specific privileges, some with read access to much of the system but only limited write access; others with global read and write access. Many combinations of access privileges are possible. A second benefit of using this model is that it fits very well with the SNMPv3 approach to security. Using it makes mapping privileges in the management system quite easy to match to privileges in managed devices.

- **Control access and communication methods to devices in the network are based on the user's identification and group**—The system should be able to determine the appropriate communication methods for each managed device.

- **Control what the user can do to view and or change data in the managed devices**—Relative to the access and communication function, the nature of the information that the management system allows the user to access on each managed device will vary. Some information may be configurable by a particular user group while other information is not. A distinction is made between the data in the management system and the data collected from managed devices because the latter, once stored on the management system, may be of greater strategic value than the information on each managed device. This greater strategic value warrants a higher level of security. One month's worth of customer utilization data on the management system is of greater strategic value than the last five minutes' utilization on one interface of a managed device.

- **Control what the user can export from the system**—The management system should exert control over the export of data, because there is no way to know if it has been properly secured once it leaves the system. For example, if you restrict someone's viewing privileges to certain customer data, you probably should allow the individual to export that customer data and no other. It's possible to create a security configuration matrix for each user and role in the system and modify it for each, but this would be time consuming to implement and complex to manage once deployed. Simpler is better in this context. One way to simplify this task is by creating groups to which users are added. All users in a particular group will have the same privileges.

- **Control time of day**—Some large networks have multiple operations centers and pass off control from one to the next, based on time of day. A member of a particular group may have one level of access while his network operation center is active and another during off hours. This is a fairly advanced notion, but one that designers of the management software should consider.

Communications Between the Management System and Devices

For the most part, security mechanisms used between the management application and the managed devices should be transparent to the user. The level of security used by the management system should be appropriate for the kind of management communication it has with managed devices. For a configuration operation, SNMPv3 PDUs might be sent using authentication, or authentication and privacy. A direct mapping does not always exist between the security methods of one device access approach and those another. One such example is the difference between a CLI-based access and SNMP. To the degree possible, the management software should attempt to do the best mapping it can to preserve as much security as possible.

Access Through Management Software versus Direct Device Access

After the management system has been configured to provide a particular set of access privileges to users and a particular set of security mechanisms for managed devices, its next critical responsibility is to close all back doors that have been inadvertently left open on the managed devices. In Figure 12.2, notice that managed devices are often configured to support direct access via methods that are not controlled by the external management software. This facility is an insurance policy for device control in the event of a management system failure, or when all alternative communications paths between management elements and a particular device fail.

Unless there is some coordination between the management system and the configuration of network devices, a particular type of SNMP or CLI access could be excluded from the management system while supported by the managed device; this should be by design, rather than accident.

The management software often has no information about security systems, outside its control, which administer direct access to devices. Such systems are common for the control of CLI-based access and, where possible, the management software security stance should be rationalized with these other systems. One way of doing this would be to have the management software access the other systems and report differences to the administrator of the management software. Terminal Access Controller Access Control System (TACACS+) is frequently used by routers to determine if a particular CLI command should be accepted from an individual. The command is received by the router and must then be verified. The router sends a request to a remote access server to determine if the command by the identified individual is authorized. TACACS+ is not an IETF standard, but it is widely used. Readers will find a good diagram and a basic introduction to TACACS+ in Chapter 16. {TACACS1] and [TACACS2] provides some additional information. RADIUS is an IETF standard with similar function described in [RFC2865].

In a sense, we can think of the service management system as replacing a remote access server whenever it is itself the source of communications. Indeed, this is how the SNMP model works. Communication between manager and managed device requires no interaction with a third device to establish whether or not the management operation is authorized. This can be an important attribute of the management system when network devices must be controlled during times when there is a difficulty in the network. Using the SNMPv3 approach, the management system has no external dependencies that could fail when parts of the network are not functioning correctly.

System Customization

One design goal for an effective service management system is to incorporate reasonable defaults for a great many parameters that are intelligently organized, so that operators can tune the system while it is in operation.

No two network environments are the same, and most management software will require customization. Some management software is so complex and flexible that it requires hours of expensive consulting service to configure it and keep it in running order. At the other extreme is software that claims it can do everything with little or no customization. Complexity increases acquisition and maintenance costs, whereas software that doesn't require customization is probably not flexible enough to meet any but the simplest of network requirements. In the end, a good service management system is a sophisticated one that requires care and feeding, like any other sophisticated distributed application.

Because no two systems or networks are alike, it is not possible to present a definitive list of configuration parameters for a service management system. We can, however, describe important areas of the system where flexibility can be helpfully designed into the system and wisely used by the people who manage it.

Network Services and Transaction Control Layers

The network services layer should be invisible to most users of the service management system. The adjustable elements of the system should include provisions for:

- **Controlling the impact of the management system on the network**—This is of greater concern when collecting large amounts of data than in configuration mode, but both operations have the potential for causing significant traffic in large networks. The network services layer should have parameters to control how many requests for information or configuration operations (e.g., SNMP SetRequest-PDUs) can be generated per second. Protocol-specific parameters should also be easily adjustable by anyone who understands the technology and is given access rights to see and modify these parameters. SNMP time out and retry values are good examples.
- **Dealing with alternative paths to managed devices and distributed components of the service management system**—A few management software packages have long had the facility to store multiple destination addresses for network devices having more than one interface. This can be a valuable tool when an interface fails or when a network failure makes it impossible to reach a device over a given interface. Storage of these alternatives is partly a problem for the data management subsystem. The network services layer may also need alternative interfaces to respond to an interface failure or for the purposes of load balancing, connection to redundant application components running on other systems, or contacting network devices through backup interfaces. Access to backup interfaces on the managed devices doesn't always involve using a different interface on the management system. It may, however, be helpful, especially in those networks with a separate network for management activity.
- **Controlling the impact of configuration or data collection on individual devices**—The management system should have controls to limit the number of requests or SET operations to a managed device per unit time.* It also must establish a limit on the total num-

---

*This is a feature that management software should have, because many current systems do not have the native smarts to deal with large transactions. It is reasonable to expect that, as more capability dedicated to the management function is put into management devices, the need for such handcrafted management application behavior will diminish. It will probably never be eliminated, however.

ber of outstanding requests for each device. Care must be taken to timestamp received data properly or coordinate the activation of configuration changes so that associated data elements don't get temporally too far apart. (A temporal drift could affect billing and other systems.) This becomes a tuning tool that, once set, is probably not adjusted very often.

- **Controlling the maximum number of outstanding transactions per device and across the network**—Related to the number of allowable outstanding SET operations or PDUs, is a higher-layer concept consistent with the notion that transactions exist at many levels. This set of parameters is designed to prevent many potentially costly transactions from stacking up in a system and then activating all at once, causing an unintended degradation in the performance of the managed device.

Database and Database Access Layer

Database and database access layer software should function primarily behind the scenes, without much involvement on the part of the people who run the system on a day-to-day basis. Because different business environments have different policies, however, the service management system should have an array of defaults and adjustments to address the following issues:

- **How often to archive**—Daily, every N hours, weekly, etc.
- **Where to archive data**—To flat files on the system, backup databases, etc.
- **Backup versions**—Should snapshots of the system be taken in case of failures? Do real-time synchronization parameters need to be set up and monitored when hot standby database technology is used?
- **Where and when to purge or aggregate the data**—Unless this is done on a regular basis, the system will begin to suffer performance problems or run low on disk space.
- **Internal performance parameters**—Any number of parameters particular to the database technology can have an impact on performance. If these can be safely adjusted, they should be accessible to the system administrators.

Analysis Modules

The details of the analysis functions are so specialized that it would be hard to list meaningful defaults and customization parameters. One approach is to design the system knowing that much of it must be adjustable, and hence avoiding hard coding even if the first release of your software does not have an interface for changing parameters.

Installation Customization

In many management software efforts, installation software is neglected until user testing is nearly ready to begin. The result is often a fragile script that does not travel well from the developer's environment to the various customer environments. These scripts are often particularly difficult because they must deal with a number of supported customer environments and include the installation of database systems. Most users of network management software are not database experts, so they require a very smart installation program that answers most of the questions for them. The installation software provided with commercial database systems is often intended to be quite generic and is not suitable for installation as part of a management application. It must be incorporated into the service management software installation program and pretty well hidden. The installation software can ask a few customization questions and then provide that data "under the covers" to the database and any other third-party software required.

Some of the flexibility that it is reasonable to expect in an installation package includes:

* **Different hardware configurations**—It *isn't* reasonable to expect a software vendor to support more than a few hardware variations and operating system revisions. Within these broad limits, the installation software should take care of any adjustments that must be made to get the system running correctly. Adjustments might include, for example, identification of disks and file systems.
* **With or without a database**—Some people will want a system bundled with a database, while others will have site licenses for the same database technology used in the product. The software should discover if there is an existing database installation to use or whether a new database should be installed. Even if the installation software is capable of discovering a database, it may not always be able to do so

in all situations. The management application software should always ask those performing the installation if there is an alternative (to that supplied with the management software) database they would like to use.

- **Network environment customization**—Most workable network management platforms adapt pretty well to the network environment of the system on which they are installed. This installation becomes more complex in a distributed system, which any effective management system is certain to be. During installation and subsequent system growth, it should be fairly easy to add distributed components throughout the network and get them added to the main service management system.

- **Disk storage limits based on expected user utilization**—It should not take five days of consulting work to determine the amount of disk space required for your management system. Asking a series of questions prior to installation will help you come up with reasonable estimates of how much disk and/or CPU you will need to run the service management system or certain parts of it.

   During the installation procedure, the software should ask questions about how many systems you want to manage, and how many program functions will be used over some specified period of time. The installation software should use this information to estimate the resources necessary for a successful installation. If the resources are not available, the system should indicate what is deficient, and by how much, or how long the system will be able to operate before an upgrade is needed.

- **Initial defaults for collection and reporting**—We've seen how useful it can be to customize data collection polling intervals, report formats, and analyses performed on data. To get the system up and running fairly quickly, it should be delivered with as much operational information as possible defaulted to carefully thought out values. Individuals performing the software installation should be able to customize at least some of the defaults to the user environment during the installation procedure.

- **Administrative user**—As part of the installation procedure, many systems require the definition of an administrator and a password for security purposes. The person performing the install may or may not be the person ultimately responsible for running the system. Basic installation defaults should be initially set with the knowledge that they will be changed as the system operates. From a security perspective, systems that have been defaulted with commonly seen user IDs and passwords should prompt the installer to change them so that it will not be subject to attack after the installation.

* **Data maintenance defaults**—Values for how often to consolidate data, where to output purged data, and the like should all be based on the input about system usage derived during the installation process. For example, if the system is expected to generate 100 MB of data per day, and 1 GB of disk space has been allocated for the application, data will have to be moved out or at least consolidated every 10 days. The installation software should provide some immediate guidance on these issues.

* **Upgrade from previous versions**—When we think about installation software, the next revision is generally not on our minds. It probably should be. Installation software should be equipped to check for any previous versions of the software and have procedures in place to migrate the data to the new installation if that is what the user requests. How much of this data is to move and where to move it to should be provided as defaults by the installation software. This can also be determined by the software as it is running, so long as the user can override the original defaults.

Some may question the value of writing the sophisticated installation software described. The fact is, if the software cannot be easily installed and made to run in a reasonable period of time, many users become frustrated with the system and simply give up. In other cases, they must purchase expensive consulting services from the vendor to make the system work, thus raising their cost of operation. So the customer pays for the development or lack thereof one way or another. They pay much more than if the vendor had made the initial investment in quality installation software. Of course some vendors have strong reasons for not making this kind of investment in high-quality installation software. First, it does take development time. Second, they make money out of the consulting services. I have been responsible for the design and architecture of a number of complex management systems, all using relational database technology. In no case has the cost for this quality installation software exceeded more than one and a half man-years—including testing. This is divided between the individual writing the majority of the installation software and all the other engineers on the project, doing what they need to do to support software that will install easily and reliably. The remainder is spent on testing in the required environments.

CHAPTER 13

External Interfaces

In this chapter, we examine the range of interfaces that a management system should provide, along with some of the technologies that could be used to implement them. Some of these interfaces have been the focus of standardization activities, whereas others are just beginning to receive attention. The areas of interest are:

- Graphic user interfaces
- Upstream interfaces
- Downstream interfaces

The history of management applications shows that even applications with wide-ranging functionality must integrate with many other software packages. This is essential to meet operational requirements in different network environments. An integrated system that performs fault, configuration, accounting, performance, and security is a system that will, in the long term, best meet the requirements of the organization. Integrated systems of the type just described are not available, especially in a multivendor, multimanagement protocol environment, and it is for this reason that external interfaces are so important. Without them, we can't construct a system that will perform all of the required functions of our service management system.

Graphic User Interfaces (GUIs)

It might appear that GUIs do not belong in a section on service management application software interfaces. They are included here because they represent one way that people can gain access to the information in a management system. More important, they are included here to make the point that the data provided to a GUI doesn't have to be structurally incompatible with, or even different from, other upstream interfaces. This is an essential point in light of the cost of implementing a GUI and defining interfaces to other upstream systems.

One other reason for considering GUI features in this section is that GUI-based systems will inevitably be compared with CLI, Web, and other management interfaces that run on managed devices. This comparison is often based on the need to integrate different systems. It is often easier to cobble together scripts that control a CLI than it is to write programs that drive a GUI or learn a proprietary programming interface that accesses data used by the GUI. This is an essential point:

GUIs are just that, for users—humans. Although CLIs are often used by humans (the experts), they are far better suited to ad hoc script creation, which we have discussed. They are different tools for different purposes.

Management software that includes a GUI has come in for significant criticism over the years. Some of the criticism was fair, but in other cases, the user interface code was unfairly blamed for deficiencies in the underlying code on which it depended. When software is tested prior to delivery, the user interface subsystem is an area that often gets a disproportionate number of bugs reported against it. A defect in the underlying code often manifests itself in the user interface. At this point, the bug has not yet been tracked down to the ultimate cause. This phenomenon of underreporting defects in the code that underlies the GUI continues once the product is shipped. End users are usually unable to make the distinction between a true GUI defect and one that manifests itself in the user interface as a result of a defect in the underlying code—nor should they have to understand these distinctions. Ideally, code is delivered to them in a working state as free from defects as possible. Excellent underlying software can facilitate the development of a good user interface and, by the same token, poorly designed software that sits below the user interface gets in the way of creating a useful interface.

Even assuming a reasonable software set functioning below a GUI, development of a good interface is still quite a challenge. We are dealing with information abstractions in a well-constructed GUI, and some of these abstractions are very hard to convey. It is difficult to imagine a more concise way of conveying abstractions to users than through a GUI. GUIs offer an excellent way for presenting complex data while hiding the details. When required, the user can expose these details at the appropriate level with a mouse click.

Having now defended GUI software, I also have to acknowledge that much of it does not help users perform their tasks. In fact, many vendors produced graphic software for marketing rather than operational purposes. The following section outlines some of the important issues related to GUI development for service management systems.

Characteristics of a Good Service Management Graphic Interface

For management software to work well in a services environment, we must develop effective graphic interfaces to those service management functions. To meet the diverse requirements of all the types of people who need to interact with the system, a variety of interfaces must be supplied. CLIs and MIB browsers are expert tools that require much more training and knowledge than it is reasonable to expect from the new operators who will have to come online if we are to scale managed services to profitable levels. Expert interfaces will be valuable in the future as expert tools primarily used for low-level debugging and single-machine fast status checking. By their nature, they cannot give composite views or perform network-wide configuration, data collection, or status monitoring functions.

If the CLI and other low-level tools don't meet the requirements of all but the most technically sophisticated users, and GUIs for management software have not been embraced, what is missing? This overview is not intended to be a general discussion of human–machine design issues, but an enumeration of some GUI design problems as they relate specifically to management software. Good user interface software for the management of networks and services will have four general characteristics: power, intuitiveness, consistency, and scalability.

GUIs are powerful.   Graphics are used to convey information used for encapsulation and abstraction. Much of the criticism leveled at GUIs is that they mimic CLIs and, in some cases, are even more difficult to use. This problem arises at least in part because GUIs were so often modeled after the extant CLI. The developer's goal was only to produce a clickable, iconic interface to what was already available in the CLI—not a powerful interface per se.

A powerful interface is one in which the user is able to accomplish a large number of tasks easily and with a minimum of interaction. An example of a powerful interface in the realm of network services would be one that separates the details of device configuration from the simple instruction to configure it to perform a number of network services. Imagine selecting a set of network objects and moving to a pull-down menu of services each should be configured to deliver. That would be a powerful interface, because it would result in many devices being configured in a variety of ways. Such interfaces might send chills down the spines of some network operators, because the potential for making big errors seems high. In fact, it is not really any greater than with a homegrown Perl script, which does the same thing. And, unlike a homegrown Perl script, our hypothetical menu is part of a generalized system that presumably has survived rigorous testing by the vendor.

An example of a GUI that lacks power is easy to find among the many common Web-based browser configuration utilities. To begin with, much potential is dissipated because the user must log in to each system to be configured. Then too, Web-based configuration tools make little attempt to hide the details of the services to be configured. The user must know the details of each and every knob in the device to correctly configure it. One of the reasons these Web-based applications are not as powerful as they might be is that they run on the managed device itself and not on a separate platform. As a result, they have fewer hardware resources to devote to a sophisticated graphic application, which would be more powerful.

GUIs are intuitive. Keystrokes, mouse clicks, and similar operations make intuitive sense to the user. Meaning is derived from consistent behavior, so that users encountering a new function in a familiar setting will be able to decipher its UI from the most general aspects and anticipate its meaning. For example, most of us are conditioned to try a double click when looking for more detailed information. Over time, certain patterns of interaction have become conventional in user interface design, at least in part because they have been found to make sense to users. A prime example of this is the basic drag and drop model, where we move items in a structured hierarchy through direct manipulation. That is, we put them where we want them. There are lots of examples of this, such as the nesting of file folders on a computer to represent the file system. To move a file from one folder to the next, we simply drag it to the new folder and release the mouse (drop it).

Consistency. If an operation that takes three clicks in one context can be done with two in another, I sometimes suggest that a client save his users some work at the expense of consistency. One of the criticisms leveled at some graphic applications is that they create too many windows on the screen and get cluttered. An easy-to-use interface attempts to balance the number of windows created versus making a single screen too complex.

GUIs scale well. There are two dimensions to the issue of scale in this context. The first is that the user interface (along with the rest of the system) must be able to process information about many systems and services at once. A great deal of management software looks very good in the demo booth. When it is deployed in large networks, however, performance often degrades to a point where the software becomes unusable. The second dimension to scale has to do with how a lot of data is represented in a meaningful way. A common example is the graphic map that represents a network hierarchy. An icon on the map might reflect the status of many devices. By selecting the icon, the user can then see the details of each device. This is a very simplistic example to make the point that the user interface must contain mechanisms that do a lot of encapsulation. It must be easy for the user to move through the data rapidly and intuitively. If users must perform many operations to get to the data they need in a large network environment, then the GUI has not scaled well.

User interface experts will be able to suggest a great many other considerations that could help developers conceptualize GUIs for service manager software. There is nothing unique about the four points just described, except that they are so rarely followed when we create user interface software.

GUI Implementation Approaches

The longstanding method used for developing management software GUIs first used C and later C++. C++ became very popular, because its object-oriented nature seemed to lend itself well to the task. More recently, two technologies, Web-based interfaces and Java interfaces, have eclipsed C-based methods. The common advantage they have over C or C++ is that they often produce visible results in far less time.

The Web. Web-based interfaces are relatively simple to construct. However, we've already seen some problems associated with Web-based configuration programs that run on managed devices as opposed to management platforms. These applications are not sufficiently powerful. If you are looking at performance data, this approach suffers from the fact that local systems only know about local data. For reports on the total amount of service delivered to a customer, you'll have to write a separate application that collects data from all the systems and integrates it for presentation. Why not perform this integration on a centralized system in the first place, since viewing the data from a single system is necessarily a part of this application?

Another essential problem with Web-based GUIs that run on managed devices is that they present a security problem. Among other concerns, they must be configured so that at least some level of restricted access to sensitive data is provided. If your system contains information that is of legitimate value to more than one organization, configuration of the Web pages and access control lists can be a large burden. We then have to multiply that burden by the number of systems in the network where the situation applies.

Web-based GUIs that execute on management systems instead of management devices are somewhat better, but still problematic in some respects. The problem is not with browser-based access per se.* The problem is with HTML-based Web pages, and even with scripts of one kind or another. To create powerful, intuitive, and easy-to-use interfaces, a generalized programming environment is required—like that provided with the Java programming language. HTML does not provide an equivalent environment.

This is not to say that the Web has no important role in GUI interfaces for service management systems. First, applications executed on remote systems that display information can be downloaded, assuming proper authorization from the centralized server. In addition, the applications can be constructed to check for new versions on the management system prior to starting each time. This ensures a smoother update process without consuming much time, except when a new version must be installed. The length of time it takes for the replacement depends on the size of the application to download, the complexity of the installation process, and the speed of the network. The last issue, speed of the network, is one reason why it is preferable to use locally installed applications instead of applets, which are downloaded from a Web page each time the user starts the program.

The Web can also play an important role in simple reporting functions, especially when those functions are accessed across the Internet by a variety of customers. These reports can be fairly rich, but tend not to be as interactive as those described earlier. They would not work as well for a real-time reporting system for example.

Java. Over the past few years, Java [JAVALANG] has grown in popularity as an implementation technology for GUIs. Unlike Web-based systems, which are often sensitive to browser and operating system differences, Java-based programs only have to contend with operating environment differences. A principle tenet of Java is *write once, run anywhere* (or at least on any compliant platform). However, differences do sometimes crop up when moving Java code from one platform to another, and these differences are important enough to require the developer to make changes. Some of these changes can be minimized with good coding practices.

---

*Many of us are familiar with the problems that differences in browsers can cause for users. This is a problem in addition to the basic issues of creating good interfaces using HTML.

Other advantages that Java-based systems have over their HTML-based counterparts include:

- **Startup speed when run as an application**—Java applets, designed to run within the confines of a Java-enabled Web browser, follow certain constraints [JAVA]. For this reason, we make a distinction in this discussion between full-fledged Java *applications* and *applets*. Our focus is on using the full capabilities of the Java language. A Java application is software installed on the target platform, whereas an applet is downloaded from a server for execution each time it is used. As a result, even when a Java applet and Java application have exactly the same functionality, the installed application will start up faster. Just how much faster depends on the loading of the applet from the server and the speed of the network.

- **Better facilities for user interaction**—The range of Java-based code from third parties for graphic navigation and presentation of all types of material is enormous. The software is generally far more dynamic than static HTML-based pages or those that have been augmented with JavaScript. The types of interfaces possible using only Java are absolutely necessary if we are to achieve our goal of placing management functions in the hands of more individuals with lesser skills and still ensure correct network operation.

- **Generalized interfaces**—A wide array of programmatic interfaces is available to the Java programmer, ranging from the Common Object Request Broker Architecture [CORBA], to database access APIs, to interfaces that enable the effective processing of XML-based data. An important advantage to generalized interfaces is that the GUI can now use the same interface that other upstream components in the management environment use, making integration of the software in a multivendor environment much easier.

- **Full object-oriented programming environment**—Java is a full featured object-oriented programming environment and offers a lot of support for the mechanisms intrinsic to our hierarchical models.

- **Powerful widgets available**—A large and growing array of third-party components (including some shareware) is useful for rapid application development. This phenomenon seems to have gained more traction in the Java environment than it ever did for C or C++.

- **Distributing more processing to end systems**—Because Java is a full programming environment with features like threads, system designers can distribute more processing to the GUI where appropriate, and improve the overall scale of the system.

Those of you who may be familiar with JavaScript probably wonder where it fits in this discussion. JavaScript can punch up pages that are essentially Web based, but was never intended to replace the full Java language. For those interested in additional details, the third edition of *JavaScript Unleashed* [JavaScript] describes limited role.

Downstream Interfaces

For our purposes, a downstream interface is one the management system uses to communicate with managed devices in the network. As we noted in Chapter 4, these interfaces have been the focus of a flurry of standards body activity over the past few years. Numerous new technologies have been created, many of which are incomplete or partially redundant. As a result of all this activity, an effective service management system must now support multiple downstream interfaces, as shown in Figure 11.3. Certainly SNMP- and CLI-based interfaces are the most common; whether these will be sufficient in the future is not known, but an effective service management system must support at least these two and all the necessary variations for each vendor.

Upstream Interfaces

An upstream interface, on the other hand, is one presented by core elements of the service management system to other cooperating software in the service management environment. Unlike their downstream counterparts, upstream interfaces have received relatively little standardization attention.

Upstream interfaces can be used to interface with:

- **Higher-level service management systems that might be closely tied to business systems that take orders**—Service provisioning systems, are an example.
- **Systems that collect data for accounting and billing purposes**—A range of products on the market today is designed to take output from management systems and use it in accounting and billing functions.

- **Network environment-specific software created by users to meet specific business needs.**

We have briefly touched on two upstream interfaces that are gaining in popularity, CORBA and XML. In a sense, these are both tools, just as a computer language is a tool. Whether one or both catches on with specific objects defined for the exchange of information between management systems remains to be seen.

CHAPTER 14

The Future of Management Software Systems

We'll have to wait a bit longer for networks that can cost-effectively support high-value services like those described earlier, or other services yet to be imagined. To the extent that management software is an enabling component in any high-value service proposition, we are far from our goal. All of the basic management technologies discussed in this book are available or in development; nevertheless, no solutions are sufficiently broad in scope except for those that have been cobbled together by enterprising network operators. Customers often complain about having to patch together software from many different vendors to approximate an effective management infrastructure. If the past is an indicator, a quick review of some of the organizations and entities that influence this technology may yield some idea of what's in store for management software in the near term.

Standards-based Activity

In the absence of a common intersection of goals and interests, it is likely that the many standards bodies involved in management technologies will continue to compete. In some cases, competition can help foster new ideas and create solutions where there were none. That said, it is more often the case that standards organizations foster their own technologies and, at best, accommodate or coexist with alternatives. IETF-developed management standards are frequently unrelated to those developed by ISO, even when there is a similarity in the general functions being standardized. This disjunction continues to afflict the process, although there are now relationships between the two bodies. Some of the good ideas developed in certain ISO standards do find their way into IETF activities, but only slowly.

The DMTF too has developed technologies that overlap with IETF standards. The best example is the Management Information Format (MIF) created by the Desktop Management Task Force (now the Distributed Management Task Force. (See the Encyclopedia of Networking [EON] and the DMTF URL in Appendix B for a good description and more details.) This technology allows operators to define data elements in much the same way they would create MIB Objects with the SMI [RFC2587] in the SNMP environment. At least from my perspective, the differences between the two are gratuitous. Concepts such as CIM, developed more recently in the DMTF, are interesting and have been incorporated in some areas of the IETF, at least to some degree, but not really integrated into the mainstream of management thinking.

In addition to the competitive efforts of the standards bodies, another problem concerns the value of their outputs. One of the hallmarks of the early IETF was that it was populated by network operators, researchers, and vendors. Over the years, the mix has narrowed, and attendees now predominantly represent vendors of network technology and occasionally a small clique of extremely knowledgeable representatives of the service providers—the experts so often referred to in this book. The impact on the quality of the standards is significant, because vendor self-interest no longer benefits from the broad user community counterweight that it used to enjoy. Indeed, the user community is more diverse than it once was.

There is, in fact, no longer a single user community. The Internet was once a fairly homogeneous community, and operator needs were once far simpler and less diverse than they are today. Commercialization of the Internet has created many legitimate constituencies whose needs must be addressed. The pertinent question is whether they should be addressed in one standards body or in many. And if many, how do we make the standards interoperate effectively, even when they overlap? So far, the answers to these questions have eluded us. Sometimes we encounter overlap even in standards developed by the same standards body. Standards bodies—it's worth reminding yourself of this from time to time—are not really monolithic entities. They reflect the conflicting interests of the membership, and this is especially true when vendors or governments are involved.

If the past is any predictor of the future, we are likely to see more of the same from the standards bodies. Placing our hope, in them for vastly improved management software is likely to lead to disappointment. Yet standards bodies are only part of the management software development equation; vendors and customers also play essential roles in determining the nature of the software available.

The Role of the Vendors

Vendors want to reduce costs and increase sales. They also believe that product distinctions make customers buy their products instead of their competitors'. To a certain extent, this is true, particularly if the distinction in question improves cost per unit of work performed or enables users to perform some new task. It is as true in the domain of management as it is for networking products. For example, I would not buy a router that does not properly interoperate with the other routers in my network. Unfortunately, there is no easy analogy in the management domain. A standardization of management functions broad enough in scope to define the same level of interoperability does not exist. As a result, some of the responsibility for our current situation does not lie with the vendors. Even so, some vendors speak as if such a full set of standards for all aspects of management (fault, configuration, accounting, performance, and security) is not in their interests, and would simply result in increased costs, impair their ability to differentiate their products, and cripple their ROI.

Vendors who want to do the right thing for their customers, in the belief that their customers will reward them with increased sales, have been thwarted in terms of management software. The standards bodies have done a poor job of creating specifications to support the goal of broad and effective management. So, if the standards bodies have not come through and the vendors won't do it by themselves, we must let customers drive the process of standardization and product development for management software.

User and Consumer Roles

Networking products and the technologies that drive them are influenced by many factors. With a few exceptions, such as government funding for the evolution of technology, products are driven by what manufacturers expect to produce the greatest profit. But customers can also drive the standardization of technologies for effective management software. These technologies may or may not come from a formal standards body. The technologies may simply be well-designed software that performs its intended function well for the customer. In this context, *customers* are not only the network engineer or the data center manager—although the latter are key influencers in purchase decisions, they are the penultimate users. The real users in this picture are the customers of the data centers and networks, from the smallest enterprise to the largest ISP. That is where the money is and the influence lies. These are the people who have the keenest interest in the smooth and effective operation of the network, because their business relies on the data that moves through it. Many network operators have a vested interest in the status quo, because any new technology is going to be disruptive and will cost money. This real threat can only be overcome with significant financial incentives.

Whatever one thinks of the technologies currently available, the evolution of network management software is not limited to any technology per se. Nor is it limited by the interesting alternatives that crop up each year, some of which we have reviewed. If you agree that significant improvements in management software are necessary to meet the challenges facing network operational staffs, and if technology is not the limiting factor, we're left with the pressing question of why new management solutions are not generally available. This book has considered answers ranging from underestimating the problem to overshooting the goal. As the onion unpeeled, we were also confronted with the fact that what is required in the IP management domain is an overarching approach not unlike that found in the TMN environment. For such an approach to take hold, vendor economics have to work and customers have to make them do. Here are some things users can do:

- **Drive vendors toward a more integrated solution**—As opposed to the patchwork of proprietary techniques and disparate standards and multiple variations of those standards in use today.
- **Demand better quality software that meets new needs and takes a broader view of management**—The people who are responsible for configuration must work cooperatively with those who monitor the network and servers. Both of these groups must work cooperatively with the billing, accounting, and sales departments. All needs must be given fair weighting to deliver a clear message to their suppliers of equipment and management software.
- **Be ready to pay for these features**—The total cost of these features includes more memory and CPU resources in managed devices and management application hardware. Users must also be prepared to pay more for products that have a significant management software investment, in the understanding that better service management means higher margin services. In short, customers have to elevate the importance of quality management software and show that they are willing to pay for it when it is available, before vendors will make the investments necessary.

Bottom Line

Vendors follow the money. If customers demand and are willing to pay for the benefits of improved management software, vendors will build it.

The trick is to know what you want. Said differently: be careful what you ask for—you just may get it. If we press vendors for each new standards-based technology that comes through the door, equipment vendors will do their best to incorporate them all, but will have little opportunity to integrate them fully and realize their promise. At the same time, third-party vendors of management software will have little time in which to develop useful applications if they have to chase every new technology that comes down the pike. Everyone's talents and resources will be spent writing software to integrate new transport or storage technologies, which do not by themselves solve the global problems associated with the management of complex networks that must support an increasing range of rapidly changing, complex services.

Participate. The best approach is to stay current with technologies as they are entertained by the standards organizations. This means not just reading about them after the standards are drafted, but actively following, and in some cases joining, the standardization process. Never were the benefits of an educated consumer more evident than in this context. An extension of this mindset would be for customers to form their own consortia to interact with vendors and standards bodies. This increases the likelihood of the customers voice being heard.

In the end, standards are only a small part of the solution. Much of the work to create an effective service management system is just good engineering. It takes time and money. What has been lacking up to this point is customer demand for integrated software and a willingness to pay for it, but the economics for the vendors and customers have not justified the kind of expenditure needed for such an effort. For this reason, the best course might be a consortium of some kind, dedicated to development of the service management software we have described. This, in combination with the new economics of the Internet, might be enough to cause the development of a worthwhile solution.

Be patient. Rapid partial solutions seldom solve real-world problems. It may take several years between the time standards are first proposed and their deployment and use in a network. It can take significantly longer before these technologies are integrated into workable product. But come they will, if the users demand them.

APPENDIX A

References

[BGP]
The Border Gateway Protocol has been in use for quite some time. Now in its fourth revision, there are a number of extensions that are in common use and others are in development. For a complete listing, search the RFC Editor's Web page for BGP:

http://www.rfc-editor.org/rfcsearch.html

The working group primarily concerned with BGP is the Inter-Domain Routing Working Group. Details of their activities and current Internet-Drafts, as well as related RFCs, can be found at:

http://www.ietf.org/html.charters/idr-charter.html

[BOILERPLATE]
The URL below points to the boilerplate referenced in this document. Check it for updates to ensure that you have referenced the most recent material.

http://www.ops.ietf.org/mib-boilerplate.html

[BOOCH]
Booch, Grady. *Object-Oriented Analysis and Design with Applications*, second edition, Reading, Massachusetts, Addison Wesley Longman, Inc. 1994, 14th printing, August 1998, pp. 13, 21, 42, 49, and 446.

[CIM]
CIM Specifications are found at:

http://www.dmtf.org/standards/cim_spec_v22/

[CMIP]
The Common Management Information Protocol is an important element of the overall specifications developed in the OSI model of management. There are a large number of specifications that are relevant. For more information see the following URL:

http://www.itu.int/rec/recommendation.asp?type=products&lang=e&
parent=T-REC-X

Pay special attention to X.710, X.711 and X.712. Readers will also be
interested in related specifications in the range of X.720 to X.754.

[COPS]

COPS usage for RSVP

http://www.ietf.org/rfc/rfc2749.txt

[COPS-PR]

COPS usage for Policy Provisioning (COPS-PR)

http://www.ietf.org/rfc/rfc3084.txt

[CORBA]

There are many useful references on CORBA. Here are a few that have
proved useful to me. Notice that the second two references are not wholly
about CORBA; they have a focus on Java with sections on how
CORBA integrates with Java.

Rosenberger, Jeremy, *Sams Teach Yourself CORBA in 14 Days*,
Sams, 1998.

Zukowski, John, *Mastering Java 2*, Sybex, 1998.

Paul J. Perrone and Venkata S.R. "Krishna," R. Chaganti, *Building
Java Enterprise Systems with J2EE: The Authoritative Solution*, Sams,
2000.

[DEN]

Information about DEN can be found at:

http://www.dmtf.org/standards/standard_den.php

[DIFFSERV]

Differentiated Services—RFC 2474 "Definition of the Differentiated Services Field (DS Field) in the IPv4 and IPv6 Headers" is a good place to
begin your investigation of Differentiated Services.

ftp://ftp.isi.edu/in-notes/rfc2474.txt

[DISMAN]

The Distributed Management Working Group of the IETF. They are
working in a number of areas that relate to event and alarm management as well as the distribution of management functions throughout a
network. The primary communication mechanism for this work is SNMP.

http://www.ietf.org/html.charters/disman-charter.html

[DMTF]

Information about the Distributed Management Task Force can be found at their homepage below:

http://www.dmtf.org/about/index.php

[DNS]

The major aspects of the Domain Name Systems first appeared in RFCs 1034 and 1035. Since that time it has been revised and extended. For a current list of RFCs and in-progress Internet-Drafts, see the DNS Extensions working group page at:

http://www.ietf.org/html.charters/dnsext-charter.html

There is also a working group for operational issues related to the DNS. Information about this working group may be found at the Domain Name Server Operations working group page:

http://www.ietf.org/html.charters/dnsop-charter.html

[EON]

Tulloch, M., *Encyclopedia of Networking*, Microsoft Press, 2000.

[EOS]

As the application of SNMP has spread to a large range of equipment and applications, the need for targeted enhancements became clear. The evolution of SNMP working group is chartered with examining short-term needs and providing solutions.

http://www.ietf.org/html.charters/eos-charter.html

[IANA]

The Internet Assigned Numbers Authority maintains a helpful Web page with a great deal of IP related information:

http://www.iana.org/

[INTSERV]

Integrated Services—as is the case with other technologies, there are many reference documents describing this technology. The pointer below is to the document "Integrated Services in the Internet Architecture: an Overview." It is a good place to begin an investigation of integrated services.

ftp://ftp.isi.edu/in-notes/rfc1633.txt

[IPSEC]

The IP Security Protocol Working Group is developing technology that can be used by systems that use IP to protect the data they transfer.

http://www.ietf.org/html.charters/ipsec-charter.html

[ITU]

ITU (International Telecommunication Union) homepage.
   http://www.itu.int/home/index.html

[JAVA]

Campione, M. and Walrath, K., *The Java™ Tutorial Second Edition Object-Oriented Programming for the Internet*, Addison-Wesley, 6th printing, July 2000.

[JAVALANG]

Arnold, Kenneth, Gosling, James, and Holmes, David, *The Java™ Programming Language,* third edition, Addison-Wesley, 2nd printing, 2000.

[JavaScript]

Wagner, Richard, Wyke, Allen, R.; *JavaScript Unleashed*, third edition, Sams Publishing, 2000.

[JONES]

Page-Jones, Meilir. *Fundamentals of Object-Oriented Design in UML*. Addison Wesley Longman, Inc., first printing, October 1991, pp. 28 and 124.

[LDAP]

A search of the RFC editor pages on LDAP revealed 38 reference documents. RFC2251—The Lightweight Directory Access Protocol (v3) is a good place to learn about the protocol. For a current listing, search the RFC editor pages at:
   http://www.rfc-editor.org/rfcsearch.html
   At the time of this writing, there are three working groups in the IETF that are working on LDAP related topics. The pointers for each of them is found on the Applications Area Working Group list on the IETF charters page:
   http://www.ietf.org/html.charters/wg-dir.html#Applications_Area
   Additional references for LDAP:
   Reed, Archie. *Implementing Directory Services*. McGraw-Hill, 2000.
   Howes, Timothy A., Smith, Mark, C., and Good, Gordon, S., *Macmillan Network Architecture and Development Series: Understanding and Deploying LDAP Directory Services*, Netscape Communications Corporation. 1999.

[M.3100]

International Telecommunication Union, ITU-T, Telecommunication Standardization Sector of the ITU, Maintenance Telecommunications

Management Network Generic Network Information Model, ITU-T Recommendation M.3100 (previously "CCITT Recommendation"), July 1995.

[M.3100-1]
International Telecommunication Union, ITU-T, Telecommunication Standardization Sector of the ITU, Series M: TMN and Network Maintenance: International Transmission Systems, Telephone Circuits, Telegraphy, Facsimile and Leased Circuits, Telecommunications Management Network Generic Network Information Model Amendment 1, ITU-T Recommendation M.3100—Amendment 1, March 1999.

[M.3200]
International Telecommunication Union, ITU-T, Telecommunication Standardization Sector of the ITU, Series M: ITU-T Rec. M.3200 (04/97) TMN Management Services and Telecommunications Managed Areas: Overview.

[M.3400]
International Telecommunication Union, ITU-T, Telecommunication Standardization Sector of the ITU, Series M: Recommendation M.3400 (02/00)— TMN management functions.

[OSI]
There are a large number of specifications that cover the Open Systems Interconnection (OSI) reference model. The page for Study Group 17 at the ITU Web site has pointers to many of the relevant documents. The URL for this Study Group is found at:
http://www.itu.int/ITU-T/studygroups/com17/index.html
From this page you can access the X series of recommendations. Here is the URL that will take you to that listing.
http://www.itu.int/rec/recommendation.asp?type=products&lang=e&parent=T-REC-X
Please pay particular attention to the X.210 through X.217 and X.217bis documents. Also note that many ITU documents are not free.

[OSPF]
As is the case with BGP, OSPF has been in use for some time. Details for this working group are found at:
http://www.ietf.org/html.charters/ospf-charter.html
That Web page also contains a full listing of relevant RFCs.

[PDP]
"A logical entity that makes policy decisions for itself or for other network elements that request such decisions." See RFC 3198.

[PEP]

"A logical entity that enforces policy decisions." See RFC 3198.

[PERL]

Perl is widely used as an implementation language in many network management environments, particularly in the role of a configuration tool. There are many excellent references and helpful tips available at the Perl home page:

http://www.perl.org/

[PM]

Waldbusser, S., Saperia, J., Hongal, H., Policy Based Management MIB. At the time of this writing, this document had not yet made it through the IETF approval process and as such has no permanent RFC number. As a result, references to specific pages in the Policy Based Management MIB document that are made may change when the PM is published in RFC form. Until that process is complete, see the SNMPCONF Working Group page that has pointers to the most current versions in Internet-draft form. Note that Internet-drafts are valid for a time period of only six months. See the [SNMPCONF] reference, pp. 5 and 15.

[POLICYTERM]

See RFC 3198 in Appendix B.

[RED]

Random Early Detection is one of a number of queue management mechanisms used by network devices to help control quality of service for packets in a network. There is an excellent Web page that has pointers to a number of papers in this area located at:

http://www.aciri.org/floyd/red.html

[RUMBAUGH]

Rumbaugh, J., Blaha, M., Premerlani, W., Eddy, F., and Lorensen, W., *Object-Oriented Modeling and Design*. Prentice Hall, 1991, p. 7.

[SMIng]

The Next Generation Structure of Management Information Working Group is working on the next generation data definition language. See the URL below for more information:

http://www.ietf.org/html.charters/sming-charter.html

[SNMP]

When we reference SNMP, we mean more than the "Simple Network Management Protocol." SNMP is the Internet Standard Management

Framework recognized in the IETF as the mechanism that protocols developed in IETF use for management. There has been a lot of contention over the configuration applicability of SNMP, which is one of the reasons that the SNMPCONF work was started. The [BOILERPLATE] reference in this appendix points to what could be termed the "core" aspects of the SNMP technology. This list is different in that it contains some of the many extant MIB Modules (and other documentation) that can be used to aid service verification and are referenced in this book. Some provide configuration capabilities while others help with advanced notification functions. This list is not all-inclusive, but gives some flavor of the range of existing SNMP-based information that has been standardized. See Appendix B which lists all cited RFCs by number for complete citations and a consolidated list.

* [RFC1850] OSPF Version 2 Management Information Base, November 1995.
* [RFC2287] Definitions of System-Level Managed Objects for Applications, February 1998.
* [RFC2564] Application Management MIB, May 1999.
* [RFC2981] Event MIB, October 2000.
* [RFC2594] Definitions of Managed Objects for WWW Services, May 1999.
* [RFC2595] Definitions of Managed Objects for Remote Ping, Traceroute, and Lookup Operations, September 2000.
* [RFC3014] Notification Log MIB. November 2000.
* [RFC3139] Requirements for Configuration Management of IP-based Networks, June 2001.
* [RFC2863] The Interface Group MIB, June 2000.

[SNMPCONF]
The Configuration Management with SNMP Working Group is working on ways that SNMP can be used to effectively perform configuration management functions.
http://www.ietf.org/html.charters/snmpconf-charter.html

[SNMPCONF-BACKGROUND]
http://www.ibr.cs.tu-bs.de/papers/policy-tr-00-02.ps.gz

[SSH]
SSH® is a registered trademark of SSH Communications Security. This technology is widely used to secure communications across the network.

It is particularly common to use this as a replacement for the telnet protocol. For further information about SSH, see:

http://www.ssh.com/

[SSL/TLS]

The Secure Socket Layer (SSL) was produced originally by Netscape. The Transport Layer Protocol version 1.0 is based on version 3.0 of SSL. The purpose of this technology is to protect communications between clients and servers. See RFC 2246 for additional information.

[TACACS1]

Velte, Toby J., Hanson, Amy and Velte, Anthony T., *Cisco Internetworking with Windows NT and 2000*, Osborne/McGraw-Hill Companies, 2000.

[TACACS2]

Burton, William, *Remote Access for Cisco Networks*, McGraw-Hill Companies, 2000.

[WANG]

Wang, Haojin, *Telecommunications Network Management*, The McGraw-Hill Companies, Inc., 1999.

[X.500]

Recommendation X.500 (02/01)—Information Technology—Open Systems Interconnection—The Directory: Overview of concepts, models, and services.

[XML]

Extensible Markup Language. For a good overview of the language, see pointers at:

http://www.w3.org/XML/

For more reading on XML:

Marchal, B., *XML by Example*, Second Edition, Que, 2002.

St. Laurent, S., Johnston, J., and Dumbill, E., *Programming Web Services with XML-RPC*, O'Reilly, 2001.

Harold, E. R., *XML Bible*, Second Edition, Hungry Minds, Inc., 2001.

APPENDIX B

RFCs by Number

This appendix is a list of RFCs that have been referenced or are relevant to topics in this book in one way or another. See the RFC Editor homepage at to access the complete set of IETF RFCs at:

http://www.rfc-editor.org/

[RFC1155]
Structure and Identification of Management Information for TCP/IP-based Internets, STD 16, Rose, M., and K. McCloghrie, May 1990.

[RFC1157]
Simple Network Management Protocol, STD 15, Case, J., Fedor, M., Schoffstall, M., and J. Davin, May 1990.

[RFC1212]
Concise MIB Definitions, STD 16, Rose, M., and K. McCloghrie, March 1991.

[RFC1215]
A Convention for Defining Traps for use with the SNMP, M. Rose, March 1991.

[RFC1270]
SNMP Communications Services, Kastenholz, F., October 1991.

[RFC1850]
OSPF Version 2 Management Information Base, Baker, F. and Coltun, R., November 1995.

[RFC1901]
Introduction to Community-based SNMPv2, Case, J., McCloghrie, K., Rose, M., and S. Waldbusser, January 1996.

[RFC1905]
Protocol Operations for Version 2 of the Simple Network Management Protocol (SNMPv2), SNMPv2 Working Group, Case, J., McCloghrie, K., Rose, M. and S. Waldbusser, January 1996.

[RFC1906]
Transport Mappings for Version 2 of the Simple Network Management Protocol (SNMPv2), Case, J., McCloghrie, K., Rose, M., and S. Waldbusser, January 1996.

[RFC1907]
Management Information Base for Version 2 of the Simple Network Management Protocol (SNMPv2), SNMPv2 Working Group, Case, J., McCloghrie, K., Rose, M. and S. Waldbusser, January 1996.

[RFC1908]
Coexistence between Version 1 and Version 2 of the Internet-standard Network Management Framework, SNMPv2 Working Group, Case, J., McCloghrie, K., Rose, M. and S. Waldbusser, January 1996.

[RFC2210]
The Use of RSVP with IETF Integrated Services, Wroclawski, J., September 1997.

[RFC2011]
SNMPv2 Management Information Base for the Internet Protocol Using SMIv2, McCloghrie, K., November 1996.

[RFC2246]
The TLS Protocol Version 1.0, Dierks, T., Allen, C., January 1999.

[RFC2287]
Definitions of System-Level Managed Objects for Applications, Krupczak, C. and Saperia, J., February 1998.

[RFC 2474]
Definition of the Differentiated Services Field (DS Field) in the IPv4 and IPv6 Headers.

[RFC2564]
Application Management MIB, Kalbfleisch, C., Krupczak, C., Presuhn, R., and Saperia, J., May 1999.

[RFC2570]
Introduction to Version 3 of the Internet-standard Network Management Framework, Case, J., Mundy, R., Partain, D., and B. Stewart, April 1999.

[RFC2571]
An Architecture for Describing SNMP Management Frameworks, Harrington, D., Presuhn, R. and B. Wijnen, April 1999.

[RFC2572]
Message Processing and Dispatching for the Simple Network Management Protocol (SNMP), Case, J., Harrington, D., Presuhn, R. and B. Wijnen, April 1999.

[RFC2573]
SNMP Applications, Levi, D. B., Meyer, P. and Stewart, B., April 1999.

[RFC2574]
User-based Security Model (USM) for version 3 of the Simple Network Management Protocol (SNMPv3), Blumenthal, U. and Wijnen, B., April 1999.

[RFC2575]
View-based Access Control Model for the Simple Network Management Protocol (SNMP), Wijnen, B., Presuhn, R. and K. McCloghrie, April 1999.

[RFC2576]
Coexistence between Version 1, Version 2, and Version 3 of the Internet-standard Network Management Framework, Frye, R., Levi, D., Routhier, S., and Wijnen, B., March 2000.

[RFC2578]
Structure of Management Information Version 2 (SMIv2), STD 58, McCloghrie, K., Perkins, D. and J. Schoenwaelder, April 1999.

[RFC2579]
Textual Conventions for SMIv2, STD 58, McCloghrie, K., Perkins, D. and J. Schoenwaelder, April 1999.

[RFC2580]
Conformance Statements for SMIv2", McCloghrie, K., Perkins, D., and Schoenwaelder, J., April 1999.

[RFC2594]
Definitions of Managed Objects for WWW Services, Hazewinkel, H., Kalbfleisch, C., and Schoenwaelder, J., May 1999.

[RFC2595]
Definitions of Managed Objects for Remote Ping, Traceroute, and Lookup Operations, S White, K., September 2000.

[RFC2748]
The COPS (Common Open Policy Service) Protocol, Boyle, R., Cohen, R., Durham, D., Herzog, S., Rajan, R., and Sastry, S., January 2000.

[RFC2749]
COPS usage for RSVP, Boyle, R., Cohen, R., Durham, D., Herzog, S., Rajan, R., and Sastry, S., January 2000.

[RFC 2753]
Levi, D., Meyer, P., Stewart, B., "SNMP Applications," April 1999.

[RFC2863]
The Interface Group MIB, McCloghrie, K.and Kastenholz, F., June 2000.

[RFC2865]
Remote Authentication Dial In User Service (RADIUS), Rigney, C., Willens, S., Rubens, A., and Simpson, W., June 2000.

[RFC2981]
Event MIB, Kavasseri, R., Stewart, B. October 2000.

[RFC3014]
Notification Log MIB, Kavasseri, R.and Stewart, B., November 2000.

[RFC3060]
Moore, B., Ellesson, E., Strassner, J., Westerinen, A., "Policy Core Information Model—Version 1 Specification," February 2001.

[RFC3084]
COPS Usage for Policy Provisioning (COPS-PR), Chan, K., Seligson, D., Gai, S., McCloghrie, K., Herzog, S., Reichmeyer, F., Yavatkar, R., Smith, A., March 2001.

[RFC3139]
Requirements for Configuration Management of IP-based Networks, McCloghrie, K., Sanchez, L., and Saperia, J., June 2001.

[RFC3198]
Terminology for Policy-Based Management, Westerinen, A., Schnizlein, J., Strassner, J., Scherling, M., Quinn, B., Herzon, S., Huynh, A., Carlson, M., Perry, J., Waldbusser, S. January 2002.

APPENDIX C

Useful URL Pointers

The following listing contains pointers to Web sites that sometimes contain a great deal of information as opposed to a single document. Some of these Web sites are also found in the alphabetic index.

CIM Specification
http://www.dmtf.org/standards/cim_spec_v22/

Distributed Management Task Force homepage
http://www.dmtf.org/about/index.php

IETF Operations and Management Area—current working groups
http://www.ietf.org/html.charters/wg-dir.html#Operations_and_Management_Area

ITU (International Telecommunication Union) homepage
http://www.itu.int/home/index.html

ITU publications page
http://www.itu.int/publications/telecom.htm

Policy Core Information Model
http://www.ietf.org/rfc/rfc3060.txt

Policy Framework Working Group
http://www.ietf.org/html.charters/policy-charter.html

SNMPv3 Working group
http://www.ietf.org/html.charters/snmpv3-charter.html

SNMPCONF Working Group
http://www.ietf.org/html.charters/snmpconf-charter.html

INDEX

Note: Boldface numbers indicate illustrations; italic f indicates that term is found in footnote.

ABOUT THE AUTHOR

JON SAPERIA (Watertown, MA) is President and founder of JDS Consulting (www.jdscons.com). He is an independent consultant in standards-based management technology for equipment vendors, ISPs, enterprises and third-party application developers. Previously he was Software Development Director at IronBridge networks and Network Architect at Digital Equipment Corp. He has extensive experience in network systems development and deployment. For the past 10 years he has actively worked in the Internet Engineering Task Force as a contributor, author, and working group chair in many areas related to SNMP based management. He has created management software product direction for systems and applications and has led architecture, design and development efforts for award-winning management software efforts.

You can reach Jon at saperia@jdscons.com.

www.ingramcontent.com/pod-product-compliance
Lightning Source LLC
Chambersburg PA
CBHW082108070326
40689CB00052B/3838